Excellence in Practice

Innovation and Excellence in Workflow and Imaging

Excellence in Practice

Innovation and Excellence in Workflow and Imaging

Layna Fischer

Future Strategies Inc., Book Division
Lighthouse Point, Florida

Excellence in Practice — Innovation and Excellence in Workflow and Imaging
Published by Future Strategies Inc., Book Division

99 98 97 1 2 3 4 5
ISBN 0-9640233-5-0

For information: Future Strategies Inc., Book Division
3116 North Federal Highway,
Lighthouse Point FL 33064 USA
954 .782 .3376
954 .782 .6365 fax
awards@waria.com

Cover design by Pearl & Associates

Library of Congress Catalog Card No. 96-83784
Publisher's Cataloging-in-Publication Data
Excellence in Practice —Innovation and Excellence in Workflow and Imaging
/Layna Fischer
p. cm.
Includes bibliographical references, glossary, appendices and index.
ISBN 0-9640233-5-0
1. Organizational Change. 2. Technological Innovation.
3. Information Resources Management. 4. Information Technology.
5. Total Quality Management. 6. Management Information systems. 7. Office Practice-Automation. 8. Imaging. 9. Workflow.
Fischer, Layna

TABLE OF CONTENTS

Gak Nederland b.v. Amsterdam, The Netherlands 67

European Excellence Awards: Workflow, Gold

Moving the Competitive Goalposts
Seven Ways to Achieve Excellence

Connie Moore, Vice President, Giga Information Group

The competitive playing field for workflow and imaging continues to change as visionary companies push the envelope for innovation and excellence. Companies excelling in document imaging and workflow share common characteristics that other organizations are well advised to learn. Companies competing head-to-head against such visionaries must shift their IT and business strategies to keep pace. Ways to move the competitive goalposts when implementing imaging and workflow technology include focusing on enterprise-wide solutions while also reaching the extended enterprise, and empowering users to develop, modify, and enhance flexible, workflow-enabled processes.

Insights on Excellence

Giga Information Group annually recognizes organizations that have demonstrably excelled in implementing innovative document and workflow solutions to meet strategic business objectives. The prestigious Giga Excellence Awards, now in their eighth year and competed on a worldwide basis, are highly coveted by organizations that seek recognition for their achievements. These awards not only provide a spotlight for companies that truly deserve recognition, but also provide tremendous insights for organizations wishing to emulate the winners' successes. There is no better way to achieve excellence than to learn from others' mistakes and successes.

To be recognized as winners, companies must address three critical areas: excellence in *innovation*, excellence in *implementation* and excellence in strategic *impact* to the organization.

- Innovation encompasses the innovative use of technology for strategic business objectives; the complexity of the underlying business process and IT architecture; the creative and successful deployment of advanced workflow and imaging concepts; and process innovations through business process reengineering and/or continuous improvements.
- Hallmarks of a successful implementation include extensive user and line management involvement in the project while successfully managing change during the implementation process. Factors impacting the level of difficulty in achieving a successful implementation include the system complexity; integration with other advanced technologies; and the scope and scale of the implementation (e.g. size, geography, inter-company processes).
- Impact is the bottom line, answering the question "what benefits does imaging and workflow deliver to the business?" Examples of potential benefits include: productivity improvements; cost savings; increased revenues; product enhancements; improved customer service; improved quality; strategic impact to the organization's mission; enabling culture change; and— most importantly— changing the company's competitive position in the market. The

visionary focus is now toward strategic benefits, in contrast to marginal cost savings and productivity enhancements.

While successes in these categories are prerequisites for winning the Giga Excellence Awards, it would reward all companies to focus on excelling in innovation, implementation and impact when installing imaging and workflow technologies. Without doing so, they will not achieve the full potential document imaging and workflow offer. Companies must recognize that implementing innovative technology is useless unless the organization has a successful implementation approach that delivers— and even surpasses— the anticipated benefits. True visionaries are not content with merely achieving benefits; they are proactively driven to raise the standard for excellence in their industry— in essence, moving the competitive goalposts.

The Seven Ways to Achieve Excellence

We at Giga Information Group are quite fortunate to have read *all* the submissions for the excellence awards— not just the Gold and Silver winners. Having combed through hundreds of submissions over the years across many countries and continents, we can clearly discern patterns in how companies achieve excellence. While not all companies share each and every characteristic, there is enough commonality to detect seven distinct paths for achieving excellence. When several of these characteristics are combined in a single installation, it often results in visionary companies moving the competitive goalposts for their industries.

First Path—Involve Users and Customers from the Very Beginning.

One hallmark of a truly excellent implementation is a high level of user involvement, not only in the design phase, but also in all phases of the project. Companies that excel in imaging and workflow understand intuitively that "users know best." For example, in the case of Trigon Blue Cross Blue Shield, user teams and project team leaders worked together to change the very nature of the customer service representative's job. By engaging users in defining how work gets done, companies can develop multifunction processes that enhance job structures, improve employee morale, and reduce employee turnover. A pattern emerges: companies excelling in workflow and imaging not only believe users should participate extensively in design activities (e.g. such as JAD, design reviews, and prototyping), but should also remain actively involved in enhancing the system following rollout.

Bank of America's Asia Division is another case in point. This business process-reengineering project focused not only on technology, but also addressed change management issues that could have otherwise derailed the project. Implementing a system across eight countries, Bank of America Asia Division concentrated on building a multi-disciplinary team, involving users extensively throughout the project, and using prototypes and pilots extensively to test ideas for the multinational implementation. Other companies must take similar steps to ensure the overall success of reengineering projects, particularly when multiple sites and multiple national cultures are involved.

Increasingly, companies will include customers in projects, in addition to involving users. Yarra Valley Water is a perfect example of a customer-centric project— customers were involved from conception to implementation and now provide ongoing inputs for continuous improvement initiatives. Yarra Valley Water was not content to focus on the internal needs of the business, but oriented the project around the *customer* and the customers' *customer* throughout the project.

Second Path—Create Flexible Work Processes

One of the biggest challenges of work automation is that processes constantly evolve as business goals, needs, and work practices change; yet modifying applications to reflect those changes takes too long. Workflow addresses this problem by associating business process rules with roles and the routing of work. Many workflow systems, however, do not allow users to modify the automated process, but rely on programmers and system developers to make changes. When this happens, workflow can become out-of-sync with real world processes, just as with an out-dated, legacy application.

Consolidated Edison and the New York City Comptrollers' Office are excellent examples of organizations that developed flexible work processes. For example, when implementing its data-centric workflow system, Consolidated Edison shifted customer service from a back office operation to a more flexible, front office process. By linking workflow to the automated call distribution system, incoming calls and letters are assigned through the wide area network to a multiskilled workforce. This allows Con Edison to manage workload peaks and valleys on a constant basis, to allocate work independently of geography, and to achieve work force flexibility on a daily and even hourly basis, while reducing customer wait time.

Third Path--Empower Users to Design and Modify Their Work Processes

The third path, empowering users to design and modify their own processes, is closely linked to the second path, creating flexible work processes. Yet many committed to the second path do not move beyond that approach by empowering users to change or adapt the automated process to their personal style. Visionaries recognize that flexibility should not only be built into the process, but also be pushed down to the users who are closest to the process.

One of the biggest work automation challenges is that business processes constantly evolve, yet modifying computer applications to reflect those changes takes too long. But as more organizations adopt case-management approaches for production work, user involvement in modifying and updating structured work processes will become increasingly more prevalent. Future implementations will be less pre-structured, more adaptable to changes in business processes and designed for users to make changes as exceptions occur and business processes evolve.

For example, although Con Edison and the NYC Office of the Comptroller require structured processes, both organizations have implemented flexible systems that allow users to modify processes. At Con Edison, "reengineering on the fly" allows customer service reps to change the workflow literally as work is processed. While at NYC Comptroller, users can access macros that adapt the workflow tasks to their personal work styles. These examples highlight that structured processes will become more flexible and adaptable to user work styles.

Fourth Path—Transcend Departmental and Geographical Boundaries

Workflow has its roots in departmental solutions. Typically "enterprise-wide" really means a big departmental solution instead of a true enterprise-wide system that transcends organizational and geographical boundaries. However, unlike more typical installations, the NYC Comptroller's Office project involved extensive BPR and workflow in multiple departments *across four separate* NYC agencies. While admittedly a challenge for many corporations, this feat must have been doubly difficult for a public agency, particularly since it involved standardizing core business processes (e.g., customer service), previously done differently within each agency.

Geographical work barriers are also being breached. For example, Con Edison has piloted a workflow project that directs incoming calls and related work items to telecommuters via ISDN lines. This allows handicapped and physically remote workers to process work exactly as if they were in the office. The visionary dimension is that the pilot supports a production process rather than the more typical ad hoc work done by telecommuters, demonstrating that the constraints of bricks and mortar can be overcome for clerical workers too.

Fifth Path—Implement Enterprise-Wide Solutions

The workflow and imaging industry has long talked about enterprise-wide implementations, but in reality those systems are very large-scale departmental solutions. But as more companies shift to cross-functional processes, enterprise-wide solutions will become a reality. Enterprise-wide processes, such as Infocamere's challenge to provide corporate records to more than 104 independent sites throughout Italy, will become more typical than atypical.

Companies wishing to gain competitive advantage through workflow and imaging must shift their focus from narrow, departmental solutions to an enterprise and extended enterprise approach. Anything less, while still generating benefits, will no longer differentiate the company, particularly as departmental solutions become more prevalent in many industries.

Sixth Path—Extend beyond the Enterprise to Reach Business Partners and Customers

Visionary companies are now using workflow and imaging to extend information and processes to customers and trading partners. For example, Capital Blue Cross has implemented a multivendor imaging system that supports health claims processing in three separate organizations (Capital Blue Cross, Pennsylvania Blue Shield and a CBC/PBS subsidiary, COMP I). Overcoming the challenges of transferring image formats from one vendor to another, Capital Blue Cross can now transfer claims from multiple mainframes across the five-site, three-company *extended* enterprise.

In a similar vein, Trigon's remote imaging and workflow satellite office at Newport News Shipbuilding supports 13,000 members. Using the same remote software as Con Edison, Trigon brings customer service to the customer's workplace, demonstrating how workflow extends the enterprise.

Yarra Valley Water, a utility in Melbourne, Australia, has recently implemented one of the most innovative, customer-centric, workflow and image-enabled business processes in the world. One of three new water retailers created through recent government restructuring, Yarra Valley Water recently installed document imaging, COLD and workflow to provide customers (law firms and plumbers) with on-line access to property sewage plans and associated mapping information. Law firms can access Yarra Valley's system via modem to submit property requests and receive property information on-line, with a money back guarantee for service levels, while independent plumbers can access Yarra Valley Water from eight plumbing supply facilities, receive information electronically without intervention, and pay on-the-spot for the service. Yarra Valley Water's vision in providing "any time, any place" customer support sets the standard for companies seeking to implement document imaging, COLD and workflow.

Seventh Path—Look Beyond Conventional Wisdom When Implementing Technology

Following conventional wisdom often leads to conventional systems that, while generating benefits, do not shift the competitive ground for an industry. Visionaries often challenge conventional wisdom, or are not content to live within its constraints. Some of the "truisms" that the 1996 award winners challenged include:

- OCR is not a viable technology for high-volume data entry;
- imaging does not work on a large scale basis because large file sizes are difficult to transfer across multiple locations;
- micrographics and optical storage do not mix; and— most importantly,
- BPR projects fail.

Martinair Holland, a Dutch airline, recently installed an imaging system to automate flight coupon retrieval. The system involves high speed scanning of flight coupons and OCR for high volume data entry. Through this installation, Martinair Holland has demonstrated that scanning items (e.g. flight coupons) and using OCR for indexing is not only viable, but can raise the bar for competitors by reducing the cost of back office ticket handling. Martinair Holland was successful in this project because careful attention was given to designing effective scanning and indexing procedures. User involvement was also key, since capture subsystems involve a significant amount of document preparation and quality assurance. By taking on the conventional wisdom, Martinair Holland set a new competitive standard within the Dutch airline industry.

Infocamere flew in the face of conventional wisdom from another perspective; it implemented a large scale, multi-site imaging system utilizing a number of image delivery mechanisms, including fax, e-mail, DAT tape, CD-R, and image transfer via wide area networks. Companies frequently shy away from large-scale, multi-site imaging installations because of concerns about network load and system performance. Companies that have this requirement, yet are reluctant to implement image delivery over WANs, should investigate alternative ways to capture and deliver images. In a very dramatic way, Infocamere showed it is possible to achieve large scale imaging in an exceedingly short time scale, and with significant benefits.

The conventional wisdom is that microfiche and optical storage do not mix—period. But to lower storage costs, Gak Netherlands, made extensive use of mixed media, including microfiche, roll film, CAR, Magneto Optical and CD-R. Since no vendor offers the comprehensive mix of storage media Gak Netherlands sought, the IT group within Gak Netherlands served as the internal systems integrator. Having gone this far, the integration group decided to integrate other third party software. For example, Gak Netherlands believed that work management would be better supported by integrating workflow with Baan's industrial logistics software, and by developing modeling tools in-house. This approach is definitely not the conventional approach for implementing imaging and workflow.

Several of the 1996 Excellence Award winners also successfully challenged the well-accepted belief that 75 percent of BPR projects fail. For example, when recent legislation shifted Gak Netherlands' customer base from 280,000 corporate customers to seven million individuals, Gak Netherlands decided to undergo extensive business process reengineering in an extremely short timescale. Not only did this project succeed, Gak Netherlands downsized the employers' insurance department by 50 percent (from 240 to 125 employees) while growing the customer base at an exponential rate. Similarly, Bank of America Asia Division's workflow and imaging project was \ by business process reengineering principles. In undertaking this project, the company developed a class for managers, IS and end users, focusing on technology concepts and reengineering principles. More than 300 individuals have taken this continuing class.

Moving the Goalposts in a Changing Competitive Environment

While today's visionaries have implemented technologies and applied the best management practices to achieve successful implementations, the competitive environment continues to change. The 1996 Gold winners— Capital Blue Cross, NYC Comptroller's, Yarra Valley Water, Gak Netherlands, and Infocamere— and the 1996 Silver winners— Trigon Blue Cross Blue Shield, Consolidated Edison, PPP healthcare, Martinair Holland and Bank of America Asia Division— represent today's most successful imaging and workflow installations.

But new projects are being planned, and even now, visionaries are in the midst of implementing systems and processes that will shift the competitive balance in their industries. Companies seeking to mirror the successes of proven winners should not merely emulate these installations, but anticipate where the document management and workflow markets are going. Future directions include migrating workflow and document management solutions to the web; expanding document management beyond text and image to incorporate video and sound; allowing customers to submit electronic documents via the internet; workflow-enabling processes between trading partners and customers; and integrating workflow with electronic commerce.

To compete successfully in the future, companies must focus on cross-functional processes that incorporate the extended enterprise. Customer service will be the single most critical battleground for both business and the public sector over the next five years. Companies can no longer sustain competitive advantage by relying on highly clerical, back-office, business-as-usual approaches to customer service. Competitive business/IT strategies must service customers *whenever* and *wherever* customers most need and value support. While the Internet will facilitate reaching out to customers, developing solid, effective processes is and will remain the most critical success factor, transcending the decision to develop customer service applications on the Web or client/server platforms. Visionary companies and government agencies will derive substantial, tangible benefits by extending workflow-enabled customer support beyond business boundaries. To do so, senior management must have the vision to drive customer-centric IT investments; business analysts must design customer-centric processes; and IT must implement technology capable of delivering on-demand customer service.

Choosing a Workflow Vendor

Most visionaries view their relationship with the vendor(s) as a true partnership. When examining the details of successful implementations, there is most likely (but not always) a highly committed vendor or team of vendors that helped make their customers' success achievable. As a result, this is a good moment to examine how the selection of a workflow vendor impacts a company's ability to achieve excellence.

Why is the vendor selection particularly important for workflow implementations? Aside from the fact that any vendor relationship is an important partnership that needs to be evaluated carefully, the workflow market will undergo major consolidation over the next 2-3 years. With more than 80-workflow vendors competing in the US market alone, and looming market consolidation, companies must select their workflow vendors carefully. Unless short-term departmental or technical needs absolutely override an organization's longer-term strategic IT requirements, corporate buyers should be wary of workflow suppliers unlikely to survive. This is particularly critical if the project is enterprise-wide or, will be implemented in phases, or if workflow is key to the company's IT strategy.

Despite the illusion created by trade publications, the market has been tough for most vendors. In fact, it has been downright treacherous: competition is keen, price declines have been steep, and workflow is both complex and time-consuming to sell.

The Seven Deadly Sins

Struggling suppliers are often guilty of the "seven deadly sins" which include:

1. Relying on systems integrators for sales; this is foolhardy-integrators are driven by specific projects rather than corporate commitments to vendors.
2. Overspecializing in a segment that is too small to be commercially viable, especially given that the demand for non-imaging workflow is still fairly low
3. Insisting on solving all workflow problems in virtually any business setting. Many vendors mistakenly pursue a "one size fits all' approach, wasting limited resources on unfocused sales efforts while losing opportunities to deepen skills.
4. Being unaware, unclear, or confused about customer requirements. For example, systems integrators' expectations differ vastly from those of user organizations, just as non-technical developers require different features than IS developers.
5. Not listening to potential customers' input, feedback, and requirements. This problem often surfaces in the demo, proposal, or site visit.

6. Trying to be both small/simple and large-scale/complex with the same product at the same time (e.g., two integrated products may be better than one "highly scalable" solution.)
7. Not recognizing that workflow is a people, process, and management problem. Some vendors fail to address critical non-technical implementation issues, putting the entire project at risk.

Workflow Supplier Evaluation Criteria

Key areas to examine include:

- Vendor's ability to meet the functional and technical requirements
- Initial and on-going costs of ownership
- Vendor's support and service policies
- Vendor's financial position
- Vendor's market position

These criteria must be examined in detail, whether the procurement involves a formal request for proposal/RFP (which is highly recommended for workflow purchases) or a short list of vendors reviewed on a less structured basis. Other crucial intangible factors include:

- Reference accounts and site visits
- A proven track record in specific vertical markets and/or applications
- Financial stability (source of capital, revenues, profits, cash flow, and debts)
- Competitive positioning in workflow
- Commitment to open platforms, the Workflow Management Coalition (WfMC) and other standards, (e.g., DMA)
- Management stability and experience
- Industry alliances and partnerships
- Related/synergistic products
- Geographic coverage
- Distribution strategy

Evaluating Workflow Vendors

Buyers should look for workflow vendors with these profiles:

- The vendor should be clearly focused on a sustainable market without deviating. (FileNet typifies a focused company; as it expands into document management, ad hoc imaging, and mass-market workflow, FileNet's market discipline is still not likely to waver.)
- Successful vendors, particularly smaller companies, are aligned with partners that enhance the vendor's competitiveness in target markets (e.g., Identitech's alliance with TSW has leveraged its presence in large utilities and manufacturing companies).
- Vertical markets are the battleground for future sales. Vendors with clear strategies are likely to succeed; conversely, small vendors without a vertical

focus will fail. (An excellent example of vertical focus is Optika's MediPower Group for medical records and patient accounting.)

- The vendor's marketing program must clearly articulate how workflow solves business problems. Workflow requires "missionary" selling, but technology-driven companies often find it difficult to convince buyers.
- Messaging is crucial; successful vendors will integrate with one or more platforms, while avoiding direct competition with Microsoft
- The Internet will be a key differentiator. Vendors that really solve the question of how to workflow-enable processes between suppliers, companies, trading partners, and customers will win big. (Action,while challenged in many areas, has taken a big step with Metro.)
- Increasingly, the integration of business process modeling tools with workflow application development tools will become a differentiator.
- Workflow is undergoing rapid change as it evolves from imaging-still the primary source of revenue. Successful vendors will strike a balance between workflow's origins and its future, while ensuring sufficient revenues in the transition.
- Successful vendors understand the difference between workflow as an *application* and workflow as a *technology* that enables applications. Only two or three workflow application vendors will survive the shakeout; the remaining successful vendors will shift to vertical solutions, packaged software, e-forms, or document management.

Technologies underpinning workflow are also changing. Successful vendors must keep abreast or ahead of the technology curve, including object-oriented development, component software, and emerging standards.

An Alternative View

If an acceptable return on investment can be achieved in a short time period, vendor credentials are less important. In these instances, the emphasis should be on selecting the right product features and functionality rather than using the vendor's long-term staying power in workflow as a knock-out criterion.

It is virtually impossible to pick the winners for a consolidating market, so rather than trying, users should focus on buying the best solution for the problem at hand. Plus, serendipity may occur; for example, a struggling vendor with an exceptional product may be acquired by a marketing juggernaut (sort of like when Wang bought Sigma). Instead of wasting time on crystal ball gazing, users should focus on identifying the requirements and matching the right product to those needs.

The Bottom Line: Findings, Conclusions and Recommendations

- The market has been tough for most vendors: competition is keen, price declines continue to be steep, and sales are both complex and time-consuming.
- Workflow tools will be incorporated into application development frameworks, middleware, e-forms products, document management products, and packaged application software.

- At the low end, workflow will rapidly shift to electronic messaging, platforms and the Internet, leaving the workflow applications market dominated by mass-market software companies.
- Vendors with clear vertical market strategies are likely to succeed; conversely, small vendors without a vertical focus will fail.
- Messaging and groupware are crucial; successful workflow vendors will integrate with one or more messaging platforms, while avoiding direct competition with Microsoft.
- The Internet will be a key differentiator for workflow vendors.
- The integration of business process modeling tools with workflow application development tools will become a differentiator.
- Successful vendors must keep abreast or ahead of the technology curve, including object-oriented development, component software, the internet and emerging standards.
- A formal request for proposal is highly recommended for workflow purchases. Key areas to examine include:
    - Vendor's ability to meet the functional and technical requirements
    - Initial and on-going costs of ownership
    - Vendor's support and service policies
    - Vendor's financial position
    - Vendor's market position
- Workflow buyers should closely evaluate prospective suppliers' long-term viability in the market, eliminating vendors with high-risk profiles from serious consideration. Factors include:
    - Reference accounts and site visits
    - Proven track record in specific vertical markets and/or applications
    - Financial stability
    - Commitment to open platforms, the Workflow Management Coalition, and other standards
    - Management stability and experience
    - Industry alliances and partnerships
    - Related/synergistic products
    - Geographic coverage
    - Distribution strategy
- Vendor credentials are particularly critical if the project is enterprise-wide, is key to the IT strategy, or involves a phased implementation.
- Buyers should not give priority to product features over vendor credentials, unless urgent business needs absolutely dictate a short-term focus and the payback is less than one year.

Conclusion

Workflow and imaging technologies have far-reaching implications. Reading the implementation case studies in this book will provide a valuable blueprint for companies trying to understand what constitutes excellence in practice. Equally as important to read, however, are the chapters that follow the case studies. These offer guidelines to process change pitfalls and insights into the perspectives of the various stakeholders in any successful implementation.

When asked what single event was the most helpful in developing the theory of relativity, Albert Einstein is reported to have answered, "Figuring out how to think about the problem."

That is an excellent way to start any process improvement project.

Connie Moore
March 1997

Arrow Trucking
Tulsa, Oklahoma

North American Excellence Awards: Imaging, Finalist

Executive Summary

By the year 2004, the U.S. Freight transportation industry will carry 11.6 billion tons of freight, generating $574 billion in revenue, according to a study commissioned by the American Trucking Association. This revenue forecast represents a 24 percent increase over 1994 revenue of $463 billion.

The Transportation Industry is a profitable and growing business. But as with all growth, there is pain. The pain that carriers feel can be summed up in two words: "Delivery Receipt." The delivery receipt starts a paper trail that has become almost uncontrollable in the transportation/shipping industry. Shippers today are demanding, and getting, a copy of the signed receipt when they get the invoice from the carrier. Or they won't pay the invoice. No proof— no payment.

Unfortunately pulling, copying and refiling the receipt is labor intensive and costly. Carriers cannot hire more clerks; the overhead drives up costs. And that is where imaging has stepped in to give relief to Arrow Trucking of Tulsa, Oklahoma.

The Solution

The Lanier solution is called CABS, or Centralized Automated Billing System. The process is simple yet effective. Arrow receives all shipping documents at the home office. These documents are scanned into Lanier's ONLINE system using a Bell & Howell rotary scanner. Clerks who assign document types to each page then index each load; delivery receipts, shipping orders, bills of lading, etc.

Billing clerks then verify the accuracy of these documents against the invoice. Once everything is accurate and all charges applied, it is ready to bill. If no documentation needs to go with the invoice, it is billed. If attachments, or copies of items to be printed with the invoice, are needed, Lanier's ONLINE system takes charge.

Through the use of Dynamic Data Exchange (DDE) with the host computer and Lanier's InterAct processor software, a special code for each shipper tells ONLINE which documents are needed. So if a particular customer wants two copies of the invoice, a delivery receipt, and a shipping order, the system pulls the invoice from the host down through the Lanier ONLINE system using COLD (Computer Output to Laser Disk) and prints the needed documents on the network printer. So each invoice is printed with the delivery receipt and delivered to the printers collated and ready to mail, saving hours and getting the bill out sooner.

"Frankly, there were two main motivations to move to imaging," says Don Battle, Vice President of Information Systems at Arrow. "One, to gain efficiency out of the process. The majority of our customers require copies of the load paperwork attached to the freight bill, and we had to find an efficient way to accommodate that problem. Two, we did it to improve service to our customers. We want to be able to bill the load correctly the first time, and if the customer has questions about that bill, we want to be able to answer those questions while they are on the phone."

Technology is not new to Arrow. Arrow is a leader in using and developing systems that make the delivery process efficient and profitable. For example, they can communicate with every driver anywhere in the United States using a satellite computer system in the cab of every truck.

The Qualcomm Global Positioning Satellite (GPS) system sends a signal every minute showing the location of every one of their trucks. They can also look at a computer screen in Tulsa and see the truck and the actual street location on a map. Lanier's ONLINE system takes advantage of this technology to signal the imaging system that a load is delivered and ready to bill.

"Qualcomm gives us the ability to route documents to the Lanier system automatically because the identifiers are already there," says Don Battle. Imaging is the best kept non secret in the transportation industry.

"We all need it," says Burt Kline owner of Atlanta Motor Lines, a $50 million LTL (Less Than Truck Load) carrier. "But this integration is nothing you lay at the feet of anybody but a major player like Lanier." Lanier's success in Transportation has not fallen on deaf ears.

"Recognizing this industry and its potential has been a focus for us at Lanier for the past two years, and we will increase our presence and offerings in transportation," says Don Savoie, Document Systems Field Marketing Director at Lanier Worldwide.

Application Overview

When a shipper calls Arrow they get a cost for shipping their particular freight. Since there can be additional costs that occur after the shipment is picked up and delivered (toll charges, special handling, etc.), the carrier cannot bill until these charges are added to the original quoted charge. These trucks are often hundreds of miles from the home office, so the driver mails the "trip package" or envelope containing all the extra charges along with his expenses.

When the trip envelope is received, all the documents are scanned into the ONLINE system and each document is assigned a document type. When all the document types necessary for billing are scanned, the status of that account is changed to "ready for billing."

Billing clerks check the accuracy of the information at the Billing terminal of the ONLINE system, and release the "load" for invoicing. Pre-assigned codes that determine which support documents need to be sent with the invoice (bill of lading, delivery receipt, delivery statement, in any combination) are coded in the Host computer (Unisys A11) and are automatically processed by Lanier's InterAct. The invoice is printed on blank Arrow invoice stock loaded in tray one of a HP IIIsi printer. Tray two holds blank paper for the required support documents. Both are compiled together as one document stack; these are picked up and mailed.

This Centralized Automated Billing (CABS) process eliminates costly billing errors, and drastically reduces the laborious and lengthy process of manually matching the invoices with support documents— a process that requires pulling files, matching documents and making trips to a copy machine.

Items of interest

6,000 invoices are sent each month. There are twelve people in the traffic department, which handle rates and three billing clerks. Arrow uses Qualcomm, a satellite system that gives them the location and status of all trucks with automatic signals coming from this computer-type device every minute. This information is displayed on PCs throughout corporation giving dispatchers instantaneous information on the status of a delivery, pickup, or an out of service truck. Qualcomm is also a great communication and safety device allowing the drivers 24-hour contact and information on their location.

Application:	CABS (Centralized Automated Billing System) with attachments
Type of carrier:	Flatbed and Specialized Carrier, and Van Division. TL, Truck Load carrier, long haul.
Equipment:	550 company-owned power units, 300 long-term lease operators, 1,000 forty-five foot flatbeds, 200 forty-eight foot flatbeds.
Terminals:	Dallas, Lone Star, and Odessa, Texas; Fontana, California; Wheatland, Pennsylvania; Ontario, Canada.
Annual Revenue:	Approximately $100 million.
System components:	10 user IMS ONLINE and IMS 2080 System.
	Three network printers (one HP4 and two HPIII si dual tray printers, one for invoice blanks and one for support documents).
	Three walk up terminals: A/P, Sales, and the Traffic Department.
	One 2137 Bell & Howell CopiScan II scanner,
	Cache Server, 32 MB, EISA motherboard, 4.2 GB Drive,
	20 user GUPTA, 88 platter 2-drive Auto Server,
	IMS 2080 COLD and an InterAct Server.

Flowchart

File folders are created in the Lanier ONLINE imaging system using Lanier InterAct preprocessor software.

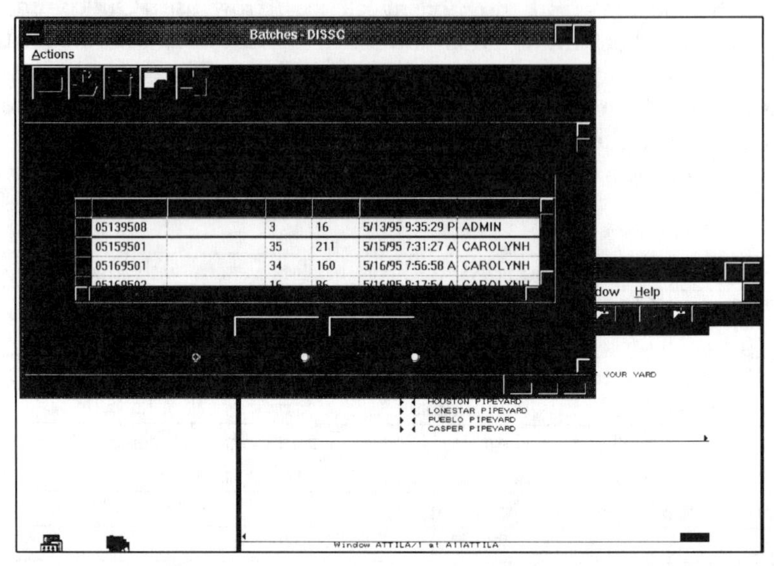

Trip envelopes (documents) arrive via mail and are scanned into the appropriate folder.

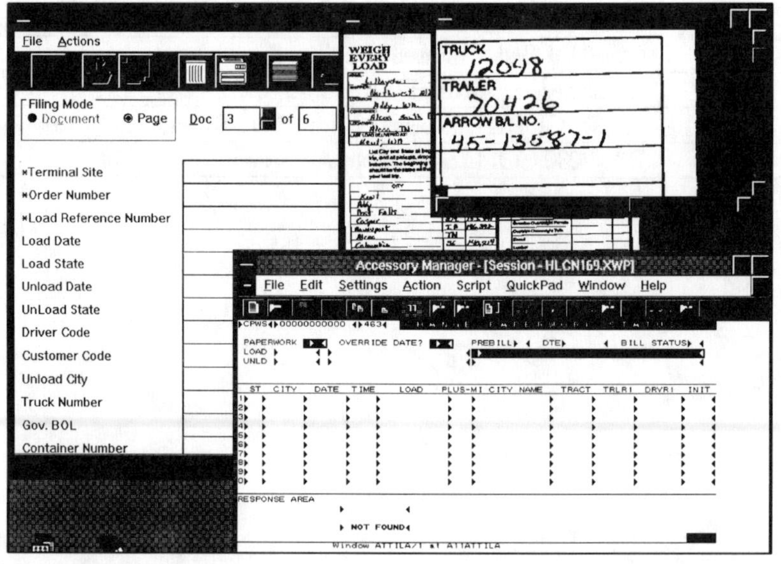

The folder field (packet Status) is updated on the host and the Lanier imaging system to reflect the receipt of the documents.

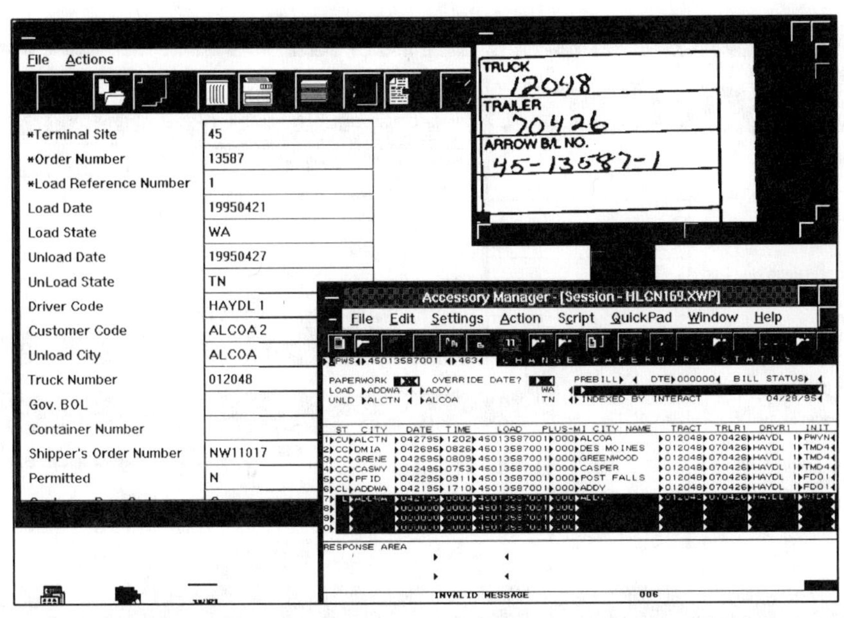

A report is generated to display or print a list of loads that are ready to be billed.

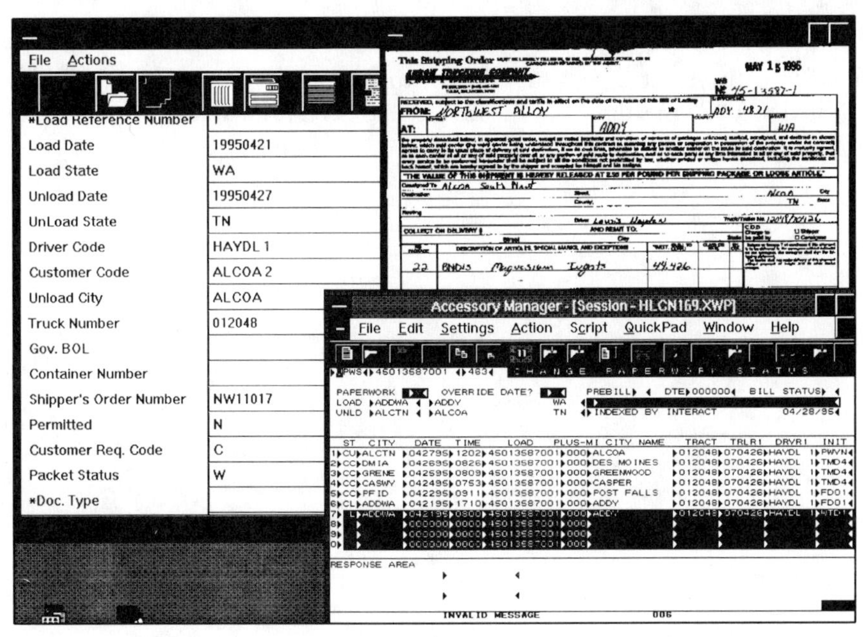

A billing clerk creates an invoice on the host system. Once the invoice is created, it is downloaded from the host to the Lanier ONLINE system (LAN).

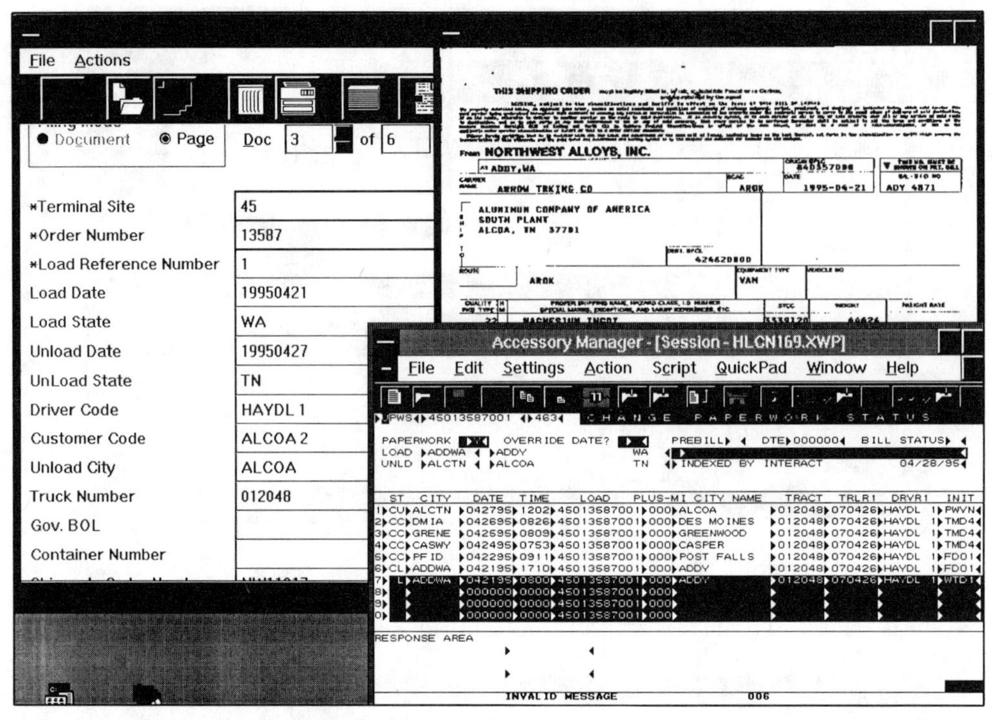

This invoice is then printed with support documents and the Packet Status Field is updated to a "P" meaning that the invoice has been printed and will not show up in the next report..

Bank of America, Asia Division
Hong Kong

Asian Excellence Award: Imaging, Silver

Executive Summary

This entry describes the successful implementation of a major workflow system in Bank of America's Asia Division. The system was implemented in a series of phases beginning in 1993 and today plays a major role in holding the line on transaction processing costs. Equally important, the workflow application has allowed Bank of America-Asia Division to continue to improve its tradition of excellent customer service.

1. Describe the system application. What is the system used for, who are the users and what does the job entail? How often or how many hours is the system in use on a daily basis?

At the end of 1995, Asia Division provided $1.6 billion (eight percent) of BankAmerica Corporation's gross income. The Division had $20.0 billion in assets and 4,300 employees. The Division had local operations in more than a dozen Asian countries.

The project began when the Division Executive Officer and the Head of Operations commissioned a task force to find ways to reduce operating costs while holding the line on customer service. The project was commissioned as the Asia Operations Center Project or, more familiarly, the AOC project.

During the 1993-planning phase of the AOC project, the project team identified two core services that could be reengineered resulting in better customer service and much lower costs. These services were very important to the success of the division and were well understood by the project team. The services were letters of credit and foreign exchange processing. Letters of credit are commercial banking documents used to transact foreign trade. They are complex legal documents and are time-critical. As would be expected, the process was paper intensive. Foreign exchange trading is at the heart of international trade, particularly in Asia.

Additionally, the project team identified one core process that was a bottleneck. This process was the delivery of messages from the central message room to the various processing departments. The message room in Hong Kong was receiving more than 3,000 messages per day. These messages had to be separated, sorted, and delivered to more than 125 workers.

When the electronic mailroom team started, their process was multi-step. As much as an hour was required to move an electronic message from the message room to user departments.

As seen on slide 1, the messages were sorted manually, entered on a control log, processed by Mail Services, and hand delivered to each user department.

Once in the department, supervisors typically resorted the messages according to work assignments. The multi-discipline team's revised process was totally electronic and now requires less than a minute. More importantly, the messages are processed in "real time" so that delivery is constant throughout the day. Messages can now be delivered directly to the worker and can be monitored centrally by the supervisor. Using the relational database behind the workflow system, each message can be tracked so manual logging is no longer necessary.

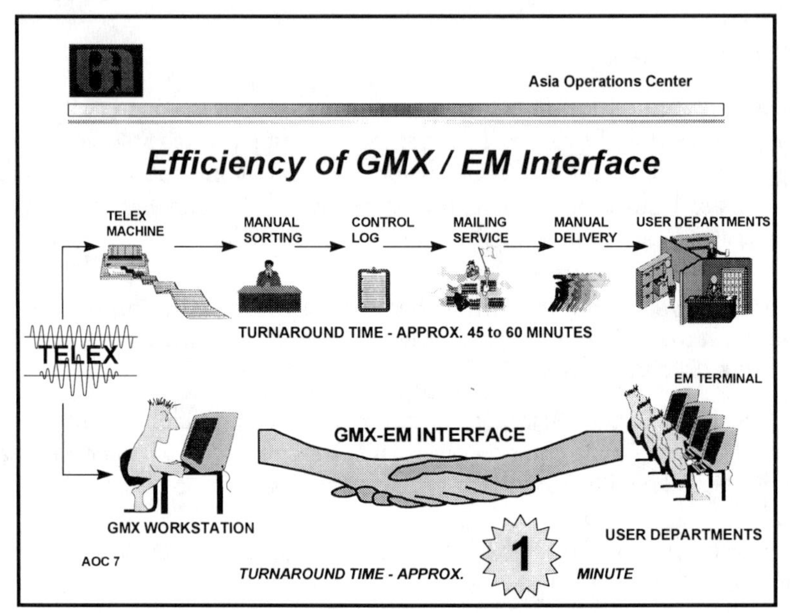

Electronic Mailroom is the heart of the AOC Workflow Application System.. Electronic messages are introduced into Electronic Mailroom via a Global Message Exchange or GMX application. GMX obtains the messages from public and private networks including Bank of America's mainframe host system - GBS. Electronic Mailroom also accepts scanned and facsimile messages.

Once in Electronic Mailroom, a program attempts to read the message. If successful, the message is committed to the optical jukebox and an entry is made on the appropriate work queue. This work queue can be anywhere on the wide area network. If the program cannot read the message, it is displayed on a "robot" personal computer referred to as Visual Sort. The Visual Sort operator is then able to add indexing information required to place the message on the appropriate work queue. Electronic messages are committed as Microsoft Word documents. Scanned and fax documents are committed as images. Because a large percentage of the traffic is electronic messages, optical storage requirements are minimal and traffic over the network is not a major issue.

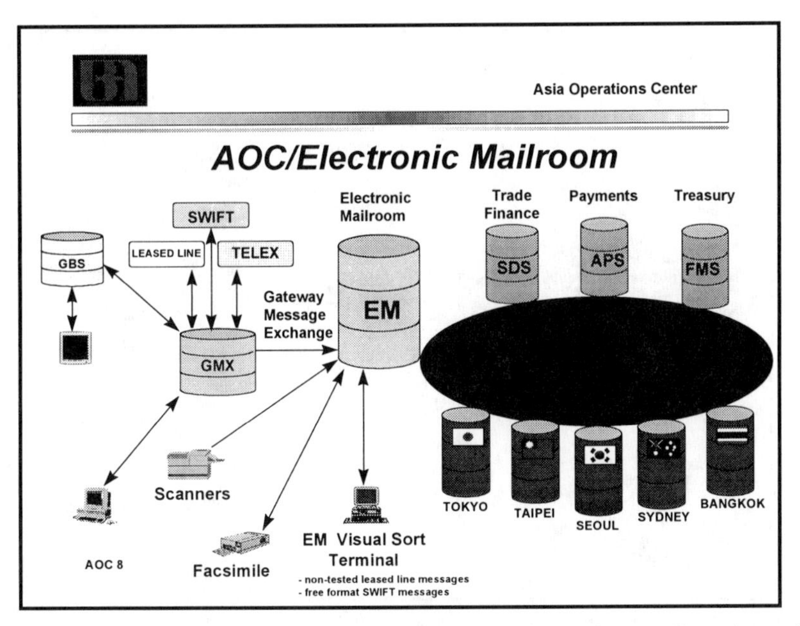

The letters of credit multi-disciplined team was able to drastically reduce the cycle time required to issue a letter of credit by using imaging and workflow technology. The letter of credit application coming in from the customer is scanned into the system. After routing through the Electronic Mailroom, it arrives in the letters of credit department and is matched with the appropriate set of host system image server using an emulator. Letter of credit issuers then construct the actual letter of credit using a PC workstation with a 21" monitor. When all approvals are received, the letter of credit is transmitted through the host system using a paperless electronic message.

During late 1994 and early 1995, other key service areas were enabled with workflow and optical. The Payments (or Remittance) function was the first key area. This was followed by additional Letter of Credit functions including the payments and negotiation processes. The letters of credit service is very paper intensive and further automation of these key functions moved the Asia Operations Center closer to a paperless environment.

2. What were the key motivations behind installing this system?

Faced with an expense curve that was increasing faster than the revenue curve, Asia Division of Bank of America set out to redesign processes as a means of slowing expense growth, without having a negative impact on customers.

3. Please describe the current system configuration: number of workstations, number and type of software, servers, scanners, printers, storage devices, number of locations involved?

The AOC infrastructure is a series of local area network (LAN) client server applications that are interconnected over a wide area network (WAN).

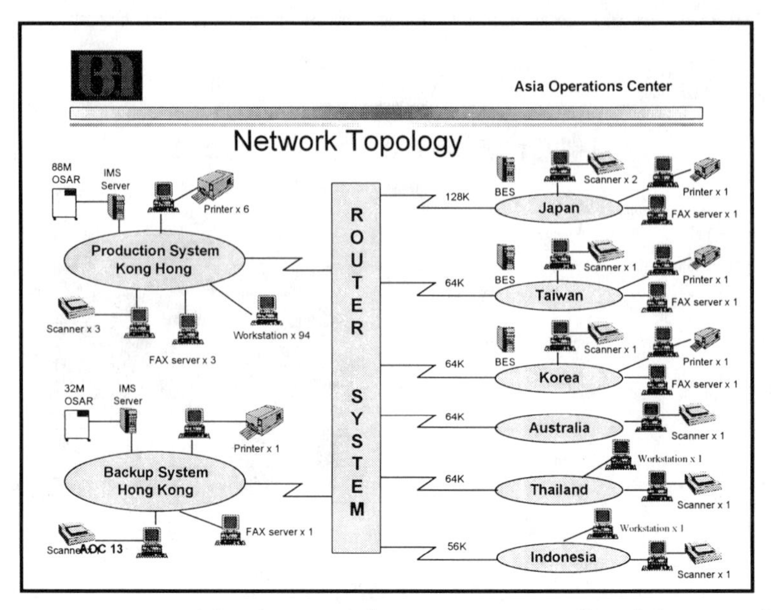

The WAN connects eight sites spread over seven countries. (The seventh country is Singapore and is not shown in this slide. Due to local data secrecy regulations, the Singapore system is standalone.) The primary system is located in Quarry Bay, Hong Kong. This system is composed of Compaq workstations and a series of FileNet servers. The main server is an IBM RISC 6000 that functions as the Optical Storage and Retrieval device as well as the FileNet workflow server. FileNet's Image Management System (IMS) supports the management of work objects stored on the optical disk device jukebox. The jukebox is a Hewlett-Packard 88mm with on-line capacity for more than four million documents.

A second configuration is housed in another building in Hong Kong about 25 minutes by subway from the primary site. This system serves as a hot backup to the system in Quarry Bay.

A series of routers connects the two Hong Kong systems with systems in Japan, Taiwan, Korea, Australia, Thailand, and Indonesia. The network is a shared network and telecommunications line capacities range from 56K to 128K.

Slide #4 provides an indication of the backup facilities of the Workflow Application System. The primary site is in Devon House, Quarry Bay, Hong Kong and the backup site in BA Tower, Central, Hong Kong. The two systems are connected via the wide area network so that the LANs can be redirected with a minimum of intervention.

The AOC workflow network is processing 5,500 transactions on a daily basis. The largest volumes are in Hong Kong, Singapore, Taipei, and Tokyo. These 5,500 images are processed on a total of 134 personal computer workstations supported by 37 servers. The servers include fax servers, scan servers, and print servers. The typical PC workstation is a Compaq 486/33 or better with 12 mb of RAM. The workstations typically have a 21" color monitor. Printing is accomplished via FileNet print servers and Hewlett-Packard LAN laser printers.

How is this system integrated with the company's other information processing systems?

An important element of the AOC system is connectivity to the host or legacy system. All of Bank of America's International branches use a single system called the Global Banking Systems or GBS. The data center for GBS is located in Croydon, outside London in England. The international units are connected via a proprietary telecommunications network. Newbridge equipment is used to connect the GBS host in Croydon to SAA gateways in Hong Kong. The gateways are, in turn, connected to the Local Area Network and can be accessed from the workstations using an emulation program.

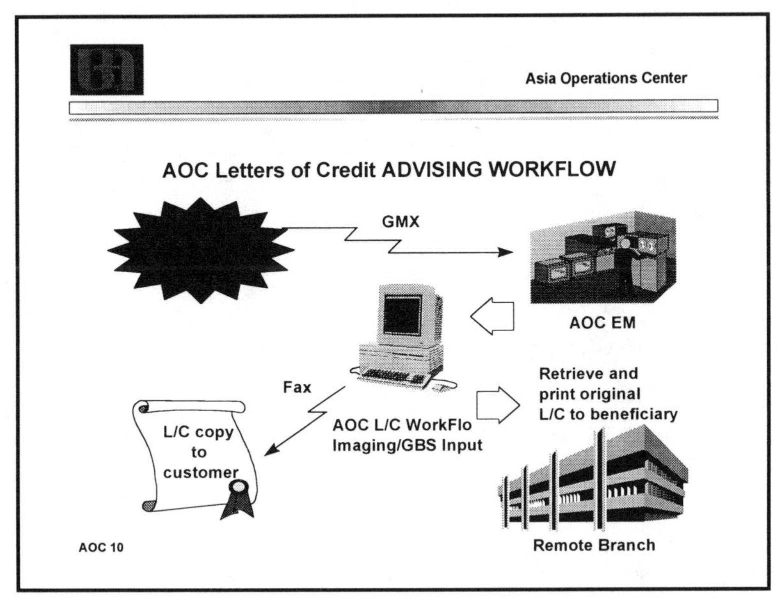

5. What stage of development is each part at; what has been installed? What is up and running? What is still in the planning stage?

The project was formally approved April 1, 1993, and the first application was implemented in August of 1993. Today, most of the workflow system's work is completed but the reengineering and consolidations continue. The Asia Operations Center has a state of the art security system driven by user profiles that determine what transactions a person can process. The Electronic Mailroom is full of efficiency features. The Foreign Exchange and Time Deposit system is maturing and operates with a minimum of interference. The Letters of Credit system is an industry leader. The newest of the systems, used to process payments, was added as part of Phase 11 of the AOC project and is also maturing.

6. Describe how the company has been impacted by this system such as:

a. What cost savings or increased revenues have been realized since the system was first installed?

The first phase of the AOC project contributed to a reduction of $1.7 million in annual operating expense. This is per annum and ongoing. Phase II resulted in a reduction of $1.3 million per annum on top of the $1.7 million saved during Phase I. Thus, the total savings per annum is currently more than $3 million.

Additionally, the project required significantly less capital expenditure and completed well under the budgeted plan. During the implementation phase, there were constant decreases in expenses related to client server technology and the project was able to capitalize on improved price performance of hardware. For example, when the jukebox was actually purchased, the cost was nearly 50 percent less than the funded amount and performance was significantly better.

Asia Operations Center

• Metrics

Country	Images	PCWS	SVRS	Total
Hong Kong	1,700	94	15	109
Taipei	890	1	3	4
Tokyo	960	1	4	5
Seoul	240	1	3	4
Sydney	220	0	1	1
Bangkok	370	1	1	2
Jakarta	68	0	1	1
Singapore	995	36	9	45
Total	5,443	134	37	171

AOC 18

b. What productivity improvements have been realized?

According to detailed records kept during the project, the net improvement in productivity for issuing letters of credit was 15.35 percent. The improvement for letters of credit advising was 15.45 percent. Most astounding, however, was the improvement in foreign exchange and money market processing which realized a 46.27 percent improvement in productivity.

During Phase II, when payments were added to the workflow system, the productivity improvement in this area was 32.66 percent.

Reduced cycle times came hand-in-hand with the improved productivity. Having the customer records on-line greatly improved control of folders and this was particularly important for letters of credit where misplaced, out of file or lost folders used to be a major effort. The customer always seemed to call about the folder that could not be found.

The ability to send faxes to customers with the click of a mouse has also been an important benefit. Customers appreciate the very fast visual responses to their questions.

Finally, although the system was supposed to eliminate paper, of course, it did not. Items that are sent to customers now, however, appear very professional as they are printed on laser printers.

c. How has the business workflow been affected (compared to before system implementation)?

Second in importance to improved customer service was improved service to the Bank of America employees. Their work environment improved significantly. Automatic logging on the PC eliminated the wide use of logbooks and registers required to track transactions.

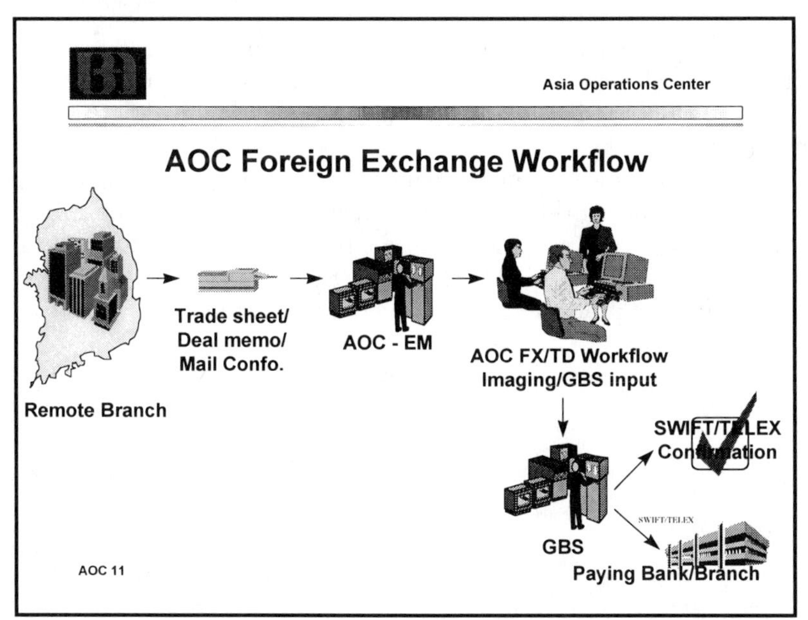

A wide variety of windows software tools can be used throughout the day to aid the daily processing that makes the Bank Americans' daily work easier and more interesting. Communication between branches is enhanced and clearer with document images used to answer questions. The offices are approaching a paperless environment and the need for manual filing is greatly reduced. With quick time to market in systems development using microcomputer tools, Bank Americans enjoy more rapid response to requests for system modifications. Some items that would have previously been changed on the mainframe system and would have taken many months are now completed in record time on the client/server system.

7. Describe the implementation process and methodology, the project team and any change in management and business process re-engineering issues addressed.

The AOC planning and development process began at the AIIM show in Chicago during April 1993. Two of the team members charged with managing the expense reduction project read Michael Hammer's book on reengineering and developed a concept that is today known as the Asia Operations Center.

The AOC project was characterized by a great deal of end-user involvement. Working with the primary software vendor, a two-week long technology concepts and reengineering principles class was developed. The curricula included workflow principles, client/server technology concepts, optical technology concepts, and reengineering principles. About two dozen employees were recruited from six Asian countries and put through the course. (The course was later tailored for different audiences and reduced to five days. More than 300 members of Asia Division management have attended this course.)

As the project moved along, considerable use was made of prototypes and pilots. One multi-disciplined team decided to centralize processing and then reengineer. Another group deemed it wiser to redefine processes, develop workflow programs, and then consolidate processing.

The Foreign Exchange multi-discipline team decided to centralize operations before streamlining the workflow so they were able to front-load a considerable amount of benefit. They realized a significant reduction in the cost of processing a transaction by using economies of scale. This was subsequently reduced even more when they reengineered processes and added workflow controls.

Budget Rent A Car, Lisle, Illinois

North American Excellence Awards: Workflow, Finalist

Executive Summary

Budget Rent A Car is a $2 billion organization with 25,000 employees operating in 117 countries and territories. It is the world's largest car and truck rental company. Budget has 3,200 corporate and licensee-owned locations, including 825 airport sites. The finance department of Budget is chartered with delivering quality services in line with Budget's values, at minimum cost, to satisfy its customers.

1. Describe the system application. What is the system used for, who are the users and what does the job entail? How often or how many hours is the system in use on a daily basis?

Budget's centralized accounts payable system is based on document-imaging and workflow software. The application is used by more than 1,000 Budget employees and is in use 11 hours a day (plus an additional seven hours a day for batch processing).

Budget chose FileNet as the system vendor for the centralized accounts payable application, and a FileNet WorkFlo/Payables™ solution was installed.

The system application follows:

Invoices are scanned and indexed, stored on optical disk and then managed by FileNet's WorkFlo application software. Features of WorkFlo/Payables include automatic invoice and purchase order matching, prioritization and electronic routing to accounts payable clerks and integration with Budget's mainframe accounts payable package by Walker Interactive Systems. Walker is a company that provides widely used host-based accounting applications.

For purchase order (PO) invoice processing, the system matches the gross amount of the invoice to the PO. If the match is successful, then WorkFlo/Payables software sends the information to the mainframe. On the mainframe, the Walker system then creates the voucher and approves the invoice for payment. With this auto-matching feature, PO invoices are paid four times faster than the manual system, eliminating the need for an accounts payable clerk to do manual retrievals or comparisons. PO exceptions are highlighted and presented to the accounts payable supervisor to monitor.

Employee travel and expense report processing are also handled with WorkFlo/Payables and the Walker system. The expense report forms are scanned into the system and index clerks key in the employee header information. Invoice clerk processors review the expense reports and enter the detail dollar information by category. The WorkFlo/Payables system generates account distributions for each line item.

For both non-PO and expense-report processing, after distributions are entered, a clerk can hit the "generate invoice key." Then, WorkFlo software writes the information to the mainframe Walker system where the invoice is placed in pending payment status for check generation. WorkFlo software is constantly checking the invoice status flag on the mainframe. If an error occurs, the invoice is brought back under FileNet WorkFlo control and sent to the clerk again.

As checks are produced on the mainframe, the pertinent information is returned to the WorkFlo/Payables system to be stored with the invoice and back-up detail for later retrieval as needed.

For checks more than $10,000, a second approval signature of a vice president is required. The executive needs to see all the backup detail along with check and invoice. With WorkFlo/Payables this documentation can be retrieved immediately at the workstation by simply entering the check number. This eliminates the clerical effort involved in manually pulling the files and matching invoices with the checks.

2. What were the key motivations behind installing this system?

In 1993, Budget had a decentralized accounts payable operation that processed more than 800,000 invoices annually and operated from more than 80 facilities. A cross-departmental team consisting of IT and finance members identified significant cost savings through a streamlined, shared-service approach that would centralize its financial administration with one consistent accounts payable system. The team also recognized that this system would:

- Increase productivity
- Increase accessibility of invoices, removing the payable functions from the regional level while leaving the regional staff with the confidence of accessibility and ultimate ownership of their expenses
- Provide a scalable solution to allow for future growth and continued improvements in other parts of Budget
- Provide a turn-key solution
- Reduce paper volume and costs, allowing staff to process by exception (process only those documents that have irregularities) and become more analytical
- Enhance the interface to the Walker Interactive accounts payable and PO systems without affecting mainframe programming

The team determined that the success of centralization was dependent upon reengineering the processes associated with payables to increase the quality of the service while decreasing the costs of providing that service.

3. Please describe the current system configuration.

FileNet WorkFlo/Payables with FileNet WorkForce Desktop[R] software on accounts payables' staff desktops. Also,

- One IBM RISC RS/6000 server
- OSAR 88M
- Two HP optical drives
- Two HP desktop scanners
- One HP 4Si MX printer attached to a FileNet print server

4. How is this system integrated with the company's other information processing systems?

The system was selected specifically for its ability to integrate with, and even enhance, Budget's mainframe and the Walker Interactive accounts payable and PO system. Details about this integration are detailed at length in Question 1.

5. Describe how the company has been impacted by this system. (Cost savings, increased revenues, productivity improvements, business workflow).

FileNet's WorkFlo/Payables and document-imaging solution helped Budget to centralize their financial systems for accounts payables. The increased efficiency in operations has meant dramatic business improvements with payback for the system beginning in a little more than one year. Specifically:

- Budget recognized a gain in productivity in the accounts payable process—increasing the number of invoices processed from 10,000 month to 65,000 month without increasing staff.
- With increased process efficiency, the cycle time for invoice processing went from five days to two days.
- Head count reduction and redeployment of 35 to 40 percent due to centralization, saving the company more than $1.1 million over the next three years.
- Centralization and increased control of the accounts payable function using WorkFlo/Payables software has enabled Budget to reduce management staff from five regional AP managers to one manager.
- Enhanced quality control. With WorkFlo/Payables there are no lost or misfiled documents or problems in locating invoices. The error rate has decreased dramatically.
- Travel and expense-reporting process is four times more efficient. One person use to handle 25 expense reports per hour, with WorkFlo/Payables an individual now can process 100 per hour.
- AP Supervisor time is used more efficiently. PO exceptions are highlighted and can be monitored by the supervisor and invoices are reviewed on exception-basis only.
- Storage space savings of approximately $12,500 per year.

Bottlenecks in workflow were eliminated by the new system because it completely restructured the way the accounts payable system works at Budget. In the previous system, the accounts payable clerk was a generalist, responsible for many things in the process. Clerks frequently were held up by supervisor delays in reviewing documents, meaning even routine financial documents could be held up in the system. With the new system, supervisors review only exceptions and each clerk is a specialist. For instance, clerks may be members of the batch prep team, the PO exception team, or the invoice processing team— allowing them to understand their work function more accurately to assess pending documents.

6. Describe the implementation process and methodology, the project team and any change in management and business process re-engineering issues addressed.

During the planning stages, Budget enlisted the help of its account payable staff to map out its work processes. Based on their feedback, Budget developed their new system. Before the system went live, Budget ran a series of acceptance tests on the system. The testing took place in modules (scanning, indexing, etc.), starting at the beginning of the process and working through each step, as each action was dependent on the one before. Several clerks participated in the testing and liked the new system so much, they pushed the IT group to implement the new system as soon as possible.

After approximately six weeks of acceptance testing, the accounts payables programs were migrated over to the new system and on June 13, 1994, the system went "live" in the manual checks and expenses divisions of accounts payable. Because the new system was so rapidly learned and embraced by the staff, Budget moved quickly to implement the system department-wide. Budget's entire financial department was operating on the new system by June 20, 1994.

Capital Blue Cross

Independent Licensee of the Blue Cross and Blue Shield Association

Harrisburg, Pennsylvania

North American Excellence Award: Gold, Imaging

Executive Summary:

Capital Blue Cross is the largest health insurer in Central Pennsylvania and the Lehigh Valley, serving nearly 1.5 million members in a 21-county region. Capital Blue Cross customers with Comprehensive Major Medical and Wraparound Major Medical coverage can be assured their claims are handled as efficiently as possible through the CBC Sigma Omnidesk system and Pennsylvania Blue Shield IBM ImagePlus system. Pennsylvania Blue Shield is also an independent licensee of the Blue Cross and Blue Shield Association. The CBC system invisibly uses automatic retrieval robots to interface with PBS and CBC mainframes and imaging systems for claims retrieval, conversion and processing among four locations. This imaging process, along with others, has reduced CBC's annual two million claims throughput by seven days.

The system is a result of a three-year, four-phase imaging implementation and internalization program across CBC to improve productivity and reduce costs. By late 1995, the high volume imaging system had completely automated over 60 percent of CBC's claims processing business. Like other imaging systems, the CBC implementation has achieved significant improvements in productivity while realizing substantial cost reductions. Specific benefits included:

- Lowered Claim Processing turnaround time by more than seven days
- Increased Claim Processing productivity by more than 25 percent
- Reduced Claims Processing costs by more than $3.5 million over five years with a two year payback
- Reduced ongoing expenses by $1.1–$1.9 million annually through staff reduction of 220 FTEs through 1998
- Improved customer service by reducing call-backs and providing quick access to information

The final CBC system distinguishes itself from other imaging systems because it represents more than a single implementation within a single organization. Instead, it demonstrates a successful, wide-ranging imaging deployment in and across organizations with unique substantial benefits, system scope, and technology innovation and implementation approaches.

- **Substantial Benefits:** In addition to claiming imaging benefits like other organizations, CBC's system has contributed to its enhanced ranking among other independent Blue Cross and Blue Shield Plans throughout the country.
- **System Scope:** The CBC system eliminates paper in a claims processing environment coordinated among *four locations, across three separate organizations, between two distinct imaging platforms, for over 1,500 CBC system users.* As such, the system

operates with processing dynamics much broader than most other imaging systems in the country. ***Imaging World*** considers this imaging implementation to be a leading example of horizontal imaging integration across separate organizations.

Technology Innovation: During its four phases, CBC introduced leading applications of imaging technology including:

- Implementation of dynamic image transfer between different image system formats
- Introduction of user-oriented production tools which augmented the capabilities of imaging
- Integration of advanced technology in scanners, jukeboxes and PCs
- Coordination of multiple mainframe systems with different imaging systems
- Use of leading workflow technologies such as priority routing, dynamic load balancing and time and event queue processing
- This innovative use of technology contributes to making CBC a leader in imaging technology application.

Implementation Approach: CBC launched its imaging program with an orientation toward building an enterprise system and internalizing support capability. This was achieved by:

- Successfully implementing a phased roll-out of imaging
- Gradually "transferring knowledge" from its integrators to CBC for full system internalization
- Effectively coordinating imaging integrators and vendor products such as The Image Consulting Group (ICG), Wang Laboratories Inc. (Wang) and IBM, to ensure proper implementation across all development phases.

This orientation toward implementation is unique because it demonstrates a complete internalization of skills, operations, and support, making the CBC's phased, internalized roll-out a model alternative approach for all major imaging implementations.

In 1996, the final CBC system captured 4,395,135 pages (supporting 1,103,167 claims) through four high-speed Kodak scanners where claims are assigned a single tracking number. From there, eight servers dynamically load-balance and route the claims to 300 claims processors who use Image Assisted Data Entry functions such as field highlighting, automated data loading and data edits to complete and check data entry.

Workstation automation assists the claims processors in checking member profiles and submitting claims to the mainframe systems for adjudication. Should a claim suspend from the adjudication process, it is routed to a Suspense Processing queue for further analysis. During suspense processing, analysts can dynamically access any claim images stored on one of CBC's four Sigma based Hewlett-Packard large capacity jukeboxes (with high-density drives). Meanwhile, any of CBC's 1,500-plus users have the ability to print or route/print images for review in supporting customer inquiries and related processing.

CBC made claim processor participation essential to the overall success and internalization of the system. From initial project kickoff in 1992, CBC maintained an Imaging Project Team and the objective that the new system for should be created with considerable involvement from CBC claims processors. As a result, the CBC Imaging

Project Team actively participated in Joint Application Development sessions, Design Reviews and Developmental Prototype Demonstrations through all phases. This approach contributed to building an understanding of system capabilities that now allows the claims processors to suggest and in many instances develop macros themselves to enhance workstation efficiencies.

CBC's Claims Processing Imaging System generates the superior productivity and cost reduction benefits which have enabled CBC to become an outstanding performer in the healthcare insurance industry. The evolution of the imaging systems also embodies the inherent scope, innovation, and approach that are elusive to many organizations implementing imaging technology today. As such, the CBC imaging system is one of only a handful of implementation stories that can offer a model approach toward a comprehensive and extensive application of imaging technology.

1.) Describe the system application. What the system is used for, who are the users and what does the job entail? How often or how many hours is the system in use on a daily basis.

Overview

The phased implementation began in early 1993 and culminated in mid-1995. Like most imaging installations, the system involves workflow and routing. In the CBC system, the focus is predominately on the high-volume processing and availability of images of health insurance claims.

Phase I: In early 1993, Wang, ICG and the CBC Imaging Project Team moved a 20-station pilot imaging environment into production to handle 200 claims per day. This area used high volume imaging viewing and routing capabilities to facilitate decision-making on claims. Once decisions were made, claims were submitted to an adjudication process on CBC's IBM mainframe. If a claim suspended because it could not be fully adjudicated, it would automatically route to a Claims Processor. Before resubmitting the claim to the host adjudication process, the claims processor would correct the claim data by viewing the image.

Phase II: By mid-1994, ICG and CBC had expanded the CBC imaging system to a production environment for 300 users and two million claims per year, 8.5 million pages. During this expansion, CBC established T1 telecommunication lines to two remote facilities and replaced all microfilm storage with imaging. CBC installed three additional Kodak high-speed scanners and four new Hewlett-Packard large capacity jukeboxes to service increased processing volumes. Six newly installed servers used Wang/Sigma's new domain architecture to appear as a single system. To provide universal access to images for all its employees, CBC provided 1,500 host terminal users with the capability to print images or route the print output to any other location.

Phase III: By spring of 1995, CBC had completely internalized and formalized imaging on the Sigma platform. Meanwhile, PBS had formalized imaging on IBM's ImagePlus platform. Each day PBS received and scanned Comprehensive Major Medical claims that had to be printed and manually shipped to the COMP I Center for processing. (Note: The COMP I Center is a joint venture of Capital Blue Cross, Pennsylvania Blue Shield and Blue

Cross of North Eastern PA.) To eliminate imaging at both locations and reduce this three-day process, CBC and PBS, with ICG, developed automatic robotics processes which transferred images on both the IBM ImagePlus System to the Wang OPEN/image™ software imaging system, in both on-line and batch modes. Therefore, claims were scanned and stored only once.

Phase IV: In the summer of 1995, as CBC built upon its internal imaging expertise, it augmented the functionality for its claims processors by introducing Image Assisted Data Entry (IADE), providing features such as field highlighting, loading and checking, to speed data entry. As part of this Phase, CBC also performed comprehensive upgrades of imaging software and hardware. In addition, operational enhancements introduced by system users were incorporated into the system.

CBC's final image system now provides claims processing throughout seven specific steps of operation: Image Capture/Indexing, Preprocess Coding, Workflow Routing, Claims Processing, Suspense Processing, Claimant Notification and Customer Service.

Image Capture and Indexing:

The final CBC imaging system captures and indexes image in two ways:

1. The PO Box number is applied to get the claims to document preparation. Users then perform pre-coding functions by indexing/sorting the claims. Certain claims have a barcode sheet attached and are routed by this barcode to specific claims processors. The majority of quality review occurs on the floor by the claims processors who send unreadable documents back to the rescan workstation for enhancements.
2. The CBC imaging system handles a large variety of incoming documents, of which 95 percent are handwritten or involve small drug receipts. In the first three phases, CBC delayed the introduction of OCR/ICR technologies because its effect on their inbound documents was limited. CBC is evaluating selective OCR/ICR preprocessing routines that can capture specific incoming data.
3. Comprehensive Major Medical claims mailed to PBS are prepared and scanned at PBS. Each night, an automated robotics process joining CBC and PBS servers, triggers the PBS mainframe to download and convert the PBS data claim record and associated ImagePlus images into batches. The images are then converted into CBC's Sigma imaging system
4. 3.Customer service representatives that require copies of comprehensive claims that were not batched to CBC overnight can retrieve these claims images directly from PBS' ImagePlus system dynamically by invoking the batch process through a single keystroke. The image is retrieved from the PBS mainframe and image system, converted and displayed into the CBC Sigma system within an average of 30 seconds using background processing. This approach replaced the previous one that took 2— 3 days.

Pre-Process Coding:

After documents are captured, they are routed to claims processors for initial data entry preprocessing. When coding, claims processors view the image and a mainframe data-entry screen. Claims processors can complete the claims entry forms much, much faster using image-assisted-data-entry (IADE), consisting of field highlights, table look-ups, field zoning/zooming and automatic field loading.

During coding, the CBC mainframe system performs Record Data Evaluation (RDE) consisting of spelling checks and logical data evaluations (i.e., provider, contract number, etc.). The system then accesses the CBC mainframe enrollment file to determine whether the customer is a valid CBC member. If a customer is not a CBC member, the claim is marked as an exception and routed to an exception queue. Any other exceptions during this validation process are flagged and handled in the same manner. Upon coding completion, mainframe fields automatically populate required image record data fields that will be used in the routing of image and data records.

Workflow Routing:

Once coded, the system establishes a priority order of claims processing. Because operational rules have been structured into exclusive hierarchies which trigger processing only in select instances, claims processing throughput is significantly faster than single level rule-based systems. During this stage, claims processors can also route claims to/from supervisors for review.

Claims Processing:

The CBC claims processors manage their work queues based upon a First-In/First-Out (FIFO) order established by the operating workflow rules. Multiple images from the client/server image system and multiple data fields from the CBC mainframe can be viewed on a single CBC image workstation. Data entry feeds both image and data records simultaneously, as required to eliminate the need for duplicate keying. If claims processors encounter a situation that requires a manager's opinion, the processor can prioritize and route the claim to a manager for immediate review. The manager can annotate the image with an opinion and return the claim to the processor.

After the claims processor finishes his/her review, the data is submitted to a Suspense Processing stage in the CBC mainframe application ESP (Electronic Suspense Processing) which matches a member's profile and history with the current claim. If, for whatever reason, the system can not process or pay the claim, the claim is suspended and forwarded to a suspend queue for processing. If the claim is payable, the mainframe system coordinates the appropriate record to notify the member.

Suspense Processing:

Suspense processors handle those claims that the CBC mainframe system identifies as exceptions. While researching the nature of the exception, suspense processors may access images, as well as annotation by the claims analyst that are associated with the image. Suspense processors can also route images and annotations for research.

Claimant Notification:

When a claim is processed successfully, the mainframe system either generates a check or prints an Explanation of Benefits (EOB) that will be mailed to the claimant. After printing, the image system removes claims from the cache to allow for faster processing. In the future, a claims processor requiring an image of a previously processed claim can retrieve it from one of the four high capacity on-line jukeboxes.

Customer Service:

CBC researches and resolves inquiries identified by mail, phone calls and faxes received from its customers. If the inquiry concerns major medical claims, customer service representatives have access to claim images and claim data screens as necessary to answer questions. Retrieval of mainframe data can automatically trigger a request to the image system to retrieve associated claim images. CBC customer service representatives use mainframe and image data to create a customer correspondence. Fax or U.S. mail can then deliver this correspondence. In addition, CBC's other 1500-plus employees operating on 3270-type terminals have the capability of printing claim images at a central site for review, or printing them remotely for other processors to review.

If customer service representatives determine that an error has occurred during processing, they can push a single key to create a new claim. They can then enter any pertinent information and route the new claim to the claims processor for adjudications. This process expedites correction and improves customer service by allowing customer service representatives to handle many problems on line that previously called for resubmitting the claim manually. This allows CBC to fully utilize the expertise and experience developed within each of the departments.

2.) What were the key motivations behind installing the system?

Providing quality service through the most efficient methods is a major goal of CBC and PBS. The National Management Information System (NMIS) creates standards of excellence in service quality and costs for independent Blue Cross and Blue Shield Plans in order to ensure consistency and quality within the Blue Cross and Blue Shield Association. Each Blue Cross and Blue Shield Plan must meet or exceed NMIS standards; and, each year the Plans are ranked by their performance against the NMIS standards. Competition is keen. Both service quality and costs must be managed equally, with little option to raise cost or lower service levels.

In 1992, CBC launched the evaluation of imaging technology to eliminate existing backlog reducing overall start-to-completion processing timeframe and address a growing concern regarding the rapid growth in major medical claims (and staff to process) that are primarily paper based, hand written and submitted by individual members. CBC had outlined a four-phase implementation approach that called for cost justification of each phase.

By mid-1995, CBC had experienced more than a 25 percent increase in operating performance. As a result, over 100 employees had been "freed-up" by imaging so they could be allocated to other operations. In addition, claims processing timeframes diminished by seven days while eliminating the need for night shift processing.

CBC was able to improve service and reduce costs effectively due to a successfully orchestrated and choreographed project approach to imaging. This involved a phased implementation of imaging technology, controlled introduction of leading edge technologies, construction of an internal development team to absorb integrators' transfer of knowledge, and the on-going education of system users to foster evolutionary enhancements to the system.

3.) Please describe the current system configuration (number and type of software, servers, scanners, printers, storage devices, etc.).

The final CBC Claims Processing System currently consists of the following configuration:

Parameters:

Sites (5):	CBC (Capital Blue Cross):	Lehigh Valley
		Harrisburg East Shore (Elmerton Ave.)
		Harrisburg West Shore (State St.)
	PBS (Penn. Blue Shield):	Pennsylvania Blue Shield Main Office
	CBC/PBS:	COMP I(Comprehensive Major Medical Claims Processing)
Image Systems:	CBC:	Wang OPEN/image (Sigma Omnidesk)
	PBS:	IBM Image Plus
Mainframes:	CBC:	IBM ES9000: CICS Application
	PBS:	IBM Multiple Mainframe Environment
Volumes:	Claims:	1,000,000 plus per year
	Claims (Comprehensive):	50,000 per year
	Pages:	8,500,000 per year
Operations:	Operators On-line:	6:30am–4:30pm Mon.–Fri.
	Operators Off-line:	8:00pm–9:00pm nightly backup and batch processing

CBC Claims System:			CBC Sites: Number		
			West	East	Lehigh
	Scanners:	Kodak 923 (upgraded from 900)	4	0	0
	Workstations:	IBM 486, 66Mhz or Pentiums	215	60	50
		16 MB RAM			
		560 MB Hard Drive			
		2 MB Winsprint 1000I Video Card			
		Windows 3.1			
		Image Print Enabled 3270 terminals	500	900	100
	Monitors:	Hitachi 20" Super Scan Color Monitor	190	60	50
		1600 x 1200 Resolution			
		72 hz scan rate			
	Image Servers:	IBM Model 95 Pentiums	8	0	0
		32 MB RAM			
		4 x 2 GB Hard Drive			
		OS/2			
		Wang OPEN/image (Sigma Omnidesk) Software			

		IBM DB2/2			
	Printer:	HP-IV SI	5	5	2
	Print Servers:	IBM PS/2 Model 90	5	5	2
		16 MB RAM			
		500 MB Hard Drive			
		OS/2			
		Wang OPEN/image (Sigma Omnidesk) Software			
	Jukebox:	Hewlett-Packard 200T	4	0	0
	Jukebox Controller:	IBM PS/2 Model 95	4	0	0
		16 MB RAM			
		1 GB Hard Drive			
	LAN:	IBM Token Ring-6 Mbit/sec	1	1	1
		OS/2 LAN Server			
	Communications:	PCOM (3270 Emulation)	1	1	1
		T1 lines to all 3 CBC sites:			

PBS & CBC COMP (Comprehensive System):			COMP	PBS
	OPEN/images Communications	T1 Line to/from PBS	1	1
	LAN:	IBM Token Ring-6Mbit/sec	1	1
	Image Transfer Robots:	IBM PS/2 Model 90	1	1
		16 MB RAM		
		500 MB Hard Drive		
		Wang (Sigma) LAN Communication Facility		

4.) How is this system integrated with the company's other information processing systems?

The CBC client/server based Claims Processing System is unique because it integrates with three major information processing environments:

- The CBC Mainframe environment including the CBC's:
  SAMM (Shared Automated Major Medical) system, including
  ESP (Electronic Suspense Processing) claims processing systems
- The PBS mainframe-based claims processing system
- The PBS mainframe-based ImagePlus image processing system

CBC Mainframe Environment:

Integration into the CBC mainframe environment took five forms:

- During *Pre-Process Coding*, "single entry" techniques were used to ensure that data input into the CBC mainframe is transferred automatically to the imaging system to eliminate dual data entry in both systems. Integration with the CBC, SAMM and ESP system is accomplished by using extensive DDE and HLLAPI interfaces to the host via Windows based terminal emulation.

- Later, during *Claims Processing*, retrieval of mainframe data screen triggers the automatic retrieval of associated images by using the mainframe data to launch an image system search and provide view-to-the workstation capability.
- After claim adjudication, the mainframe ESP system passes exception claims that have "suspended" back to the suspense processors for *Suspense Processing*. During that step, suspense processors can navigate to appropriate host screens and view claim images together to access all data necessary to research the claim. Data received from the host is used to route the record and image to the suspense processor for exception resolution.
- When performing *Customer Service*, representatives use mainframe data queries to launch retrievals of corresponding images (when desired) in order to more quickly respond to customer inquiries.
- Through the *Image Printing* function, CBC marries mainframe host terminals to the image client-server environment to allow all PC and Terminal users the ability to print. This function also provides users the ability to quickly retrieve hard copy. Printers are strategically placed throughout the CBC locations. Each department in CBC is assigned to a shared printer. By selecting a print request, users can default to their own printer or route the print to a different location for someone else to review. If both users are image enabled, this routing is done on-line without requiring production of a hard copy.

PBS Mainframe Environment:

Each day, comprehensive major medical claims located at PBS need to be directed to COMP I for review. The Comprehensive Claim Robot issues a query to the PBS Comprehensive Major Medical Claims System to transfer all relevant Comprehensive Major Medical Claims. The PBS Robot, constructed by CBC, must query the PBS mainframe for comprehensive Major Medical claim records and ready these records for transfer. The data record is then coordinated with the image transfer from PBS' IBM ImagePlus System (see description below).

PBS ImagePlus Imaging System:

As PBS mainframe records are sent to the COMP I subsidiary located at CBC as an overnight batch, the CBC robot must also query and receive the associated PBS comprehensive Major Medical claim images. To accomplish this, the CBC robot creates an empty Wang (Sigma) document and triggers a query to the PBS Robot. The PBS-link Robot obtains the PBS ImagePlus images and converts them from a MODCA-IOCA format to a TIFF Group III format for file transfer over telephone lines. CBC's use of Sigma's network management capabilities to manage image transfer represents the first application of this technology for Wang (Sigma).

Once received, a CBC server converts the TIFF Group III format to a TIFF Group IV format for input into the Wang (Sigma) format that can be used by the COMP I claims processor the next day. Because PBS claims are stored as images on the PBS ImagePlus system, and are accessible at anytime by CBC, they are maintained only on magnetic storage at CBC.

COMP I service representatives who require copies of comprehensive Major Medical claims which were not batched to CBC overnight can retrieve these claim images directly from PBS' ImagePlus system by dynamically invoking the process through a single keystroke. The image is retrieved from the PBS mainframe and image system, converted, and displayed onto the CBC image system within an average 30 seconds using background processing instead of 2— 3 days.

5.) Describe how the company has been impacted by this system. Be as specific as possible. What cost savings or increased revenues have been realized since the system was first installed?

Specific cost savings derived from the implementation of the Claims Processing Systems aggregate to more than $3.5 million over a five-year period, which corresponds to a two-year payback. In addition, CBC expects to realize an additional $1.1–$1.9 million in savings each year. These savings were generated by:

- Reducing claims processing staff and support personnel by 80–110 employees
- Reducing the document preparation, batching and filing staff by 20-40 full time employees
- Eliminating microfilm, staff and servicing costs; calculator and calculating costs.

What productivity improvements have been realized?

Specific productivity improvements realized an increase of more than 35 percent in productivity as a result from the imaging system. Improvements include:

- Eliminating all existing backlog
- Reducing claims processing time by a minimum of seven days
- Reducing access to PBS claims from three days to 30 seconds
- Enabling customer service representatives to answer a question on the initial call
- Reporting statistics on performance to restructure work profiles for optimum effectiveness:
    - Custom system reports included as part of the Claims Processing Imaging System provide CBC supervisors and management personnel with new levels of critical data that indicate how effectively CBC is processing the time-sensitive stream of ongoing claims. Comparing all scanned data and processing reports from the image system and comparing them to what was actually processed on the mainframe has developed a comprehensive reconciliation process. This is a double check to ensure accurate claim processing.
    - Supplying imaging users with tools that allow users to build on their own productivity-enhancing system features:
    - Once trained on the Norton Macro-Builder and Artist Graphics Control programs, users started creating their own productivity enhancements. A user group was formed across departments to propagate the transfer of information throughout the organization. Research functions that previously took several minutes were reduced to seconds by driving which mainframe screen appeared and how long it displayed before bring up the next.
    - The most exciting impact of the Claims Processing Imaging System is the effect that it has had on the CBC claims processing personnel. New

imaging processes were designed and approved by people who performed the daily manual processes (see implementation approach) before any application development occurred, making way for new attitudes and job satisfaction levels.

How has business workflow been affected (compared to before system implementation)?

Much of CBC's productivity increase is attributable to changes in the claims processing workflow:

- *Automated* processing in coding claims processing and comprehensive claims processing eliminated many or all manual tasks within each function. In comprehensive claims processing, for example, two "robots" automated the entire manual three-day process of getting claims from PBS to CBC.
- *Access* to a variety of systems for information (multiple mainframes and other image systems) helped to coordinate information on the claims processors desktop. This ready access to information reduced long, drawn-out process of data collection, permitting the CBC claims analysts to concentrate on specialized customer services without being disrupted by the data entry or data manipulation processes.
- *Distribution* and management of workload among multiple sites allowed CBC to dynamically balance workload processing among all its areas while centralizing system management. For example, customer service representative can now reroute errors directly to claims processors rather than resubmitting the claim for data entry.
- *Ongoing* user education and involvement contributed to the initial project success and continued effectiveness. Armed with a thorough understanding of imaging and assistance from system administrators, system users have automated most of their processes. The success which users feel from their new image-based workflow has fostered a continuous effort to identify opportunities for improvement and use of productivity tools to build personal operation macros (see user tools in Productivity Benefit section).

Other forms of impact:

The final CBC system distinguishes itself from other imaging systems because its organizational impact goes beyond productivity improvements and cost savings.

- Wide ranging *system scope* gives access to many types of documents among five locations, across three separate organizations, and between two distinct imaging platforms for over 1,500 system users.
- *Innovative technologies* not only improved productivity, but also fostered ideas for improved workflow including: dynamic transfer between different imaging systems, user development tools, latest hardware/software, access to multiple mainframe and imaging systems, and innovative workflow functions such as priority routing, dynamic load balancing and time and event queue processing.
- The day-one *approach toward implementation* that called for building toward an enterprise system and internalizing support capability was planned through a phased rollout of imaging. This allowed for continual training of users so that they can make simple system enhancements themselves. Effective coordina-

tion among the outside system integrators throughout the phases led to gradual "transfer of knowledge" that allowed CBC to fully internalize the system.

6.) Describe the implementation process and methodology, the Imaging Project Team, and any change management and business process reengineering addressed.

The CBC implementation evolved through four phases over three years and across three organizations. As such, the implementation methodology drove not only the vertical growth of imaging within CBC, but also the horizontal deployment and coordination of CBC imaging with other organizations such as PBS and COMP I.

CBC's approach combined user feedback with design analysis to determine system requirements and impact on employees. In addition, CBC validated appropriate hardware and software platform selections and determined feasible and cost effective uses for the proposed technologies. Elements of (1) Building a Project Team, (2) Managing Change and (3) Applying Reengineering were contained within the overall implementation seven-stage methodology that CBC adopted.

STAGE 1: Project Initiation and Management (Project Team/Change Management):

During Stage 1, CBC launched an initiative that was critical to the success of the final CBC system. From the beginning, CBC established an Imaging Project Team consisting of CBC technical and user personnel, as well as integration consultants. CBC established an Executive Project Sponsor for overall project accountability and a Project Manager to manage on an ongoing, day-to-day basis. Integrators working with CBC were expected to mirror this structure with their own personnel. In this way, CBC constructed a management and delivery structure that could handle all project concerns, from implementation to funding and managing project change orders, including issue escalation and resolution. User involvement was critical because it provided a practical business orientation throughout all phases.

- *Image/Workflow Project Education:* During the initial phases, CBC staff gained knowledge and experience through joint participation with ICG and Wang consultants in the delivery of the Phase I and Phase II system. In later phases, CBC maintained a Imaging Project Team to iteratively continue development with CBC systems personnel and processors. ICG participated in an advisory role.
- *Technology Education:* As the implementation progressed, it was important to communicate the capabilities of emerging technologies that the CBC users could exploit once the system was installed. This was accomplished through formal presentations.
- *Post Implementation Review:* To address unanticipated expectations and issues that appeared after installation. CBC setup a review group composed of users, system personnel and integrators. Together, this team identified and provisioned for exceptions. The involvement of decision-making staff from all areas made resolution effective and timely. As issues were resolved, meetings were reduced. The group still met as required, however, to discuss problems and share solutions.

STAGE 2: Project Analysis (Reengineering)

In Stage 2 of all CBC project phases, CBC perform Project Analysis to determine the appropriateness and cost-effectiveness of implementing new imaging/workflow technologies to support their business objectives:

- *Imaging Application Study:* In the first phase of the project, the CBC Image Project Team identified how the CBC workflows would benefit from imaging and workflow technology. The Imaging Project Team produced a Cost-Benefit Analysis that profiled costs to support these high-level flows and the resulting benefits. The Imaging Project Team also determined intangible business and system strategies, guidelines and standards that would evolve from the project.
- *Process Reengineering:* As the costs of the technologies were justified, CBC proceeded with formal reengineering of business sub-processes within the targeted CBC area. Reengineering took a more exact look at the profiled revised flows created in the Imaging Application Study and included the following steps:
    - Developing a thorough understanding of current processing flows and their rationale
    - Dividing workflow and goals into processes
    - Subdividing processes into tasks
    - Identifying essential tasks and prioritizing those tasks
    - Rebuilding processes with priority tasks while leveraging imaging/workflow technologies
    - Formalized revised workflows
    - Modeling revised workflows to determine accuracy and true benefit of reengineering
    - Making adjustments and iteratively remodeling until reengineering proved effective
    - Finalizing the reengineered workflow and building a proposal for management
    - Building a preliminary/draft implementation project plan
    - Obtaining management review/approval

STAGE 3: Project Plan and Requirement Development

After demonstrating how imaging added true value to the CBC processing environment in each phase, the Imaging Project Team embarked on building a project plan and identifying requirements for each system implementation:

- *Project Plan:* Using the draft image project plan developed in the final phase, the Imaging Project Team worked with CBC management to fine-tune project details into an executable final project plan.
- *Functional Specification:* Working with the user groups in the appropriate business areas for each CBC phase, the CBC Imaging Project Team defined user requirements that would become functional specifications for the system.

This involved the identification of workflow rules, queuing algorithms, time and event triggers and other processing strategies. In addition, the Imaging Project Team focused on reports, screen layouts, indexing and database element definition along with high-level operational procedures.

- *Navigational Prototype:* This assisted in determining functional Specifications and Project Plan in detail to ensure that initial system benefits were still intact. As such, the CBC Imaging Project Team moved toward developing detailed system design specifications.

STAGE 4: Planning and Design

Before proceeding to Stage 4, CBC management reviewed the Functional Specifications and Project Plan in detail to ensure that initial system benefits were still intact.

- *Strategic Image Systems Plan:* As part of this stage, the Imaging Project Team revisited the corporate CBC vision for the strategic, enterprise-wide roll out of imaging technology to ensure that the current phase of development was consistent with this plan. Through user education and participation the CBC created a universal interest and desire for the imaging system. Because the departments had to cost-justify their use of the new system, they participated heavily in requirements and design stages to ensure appropriate system application. This participation contributed to total buy-in to the system by many different departments.
- *System Design:* After the implementation platform was determined from the preliminary designs, the CBC Imaging Project Team produced Final System Design Specifications of the imaging and workflow system. System prototyping throughout the Systems Design stage helped expedite the design stage.

STAGE 5: Installation, Development and Testing

Before proceeding to Stage 5, CBC management reviewed the System Design again to ensure that initial system benefits were still intact. With the benefits intact, the CBC Imaging Project Team moved toward installing, developing and testing the system application:

- *System Installation:* The CBC Imaging Project Team and integrators installed basic image system hardware and software. This system provided a base platform for development, backfile conversion and training.
- *Application Development:* In CBC Phases I and II, ICG developed the CBC workflow and imaging functionality on the Wang imaging platform. By Phase III and IV, CBC had internalized this capability and used ICG, Wang and IBM in an advisory role.
- *Testing:* After application development, the Imaging Project Team installed the application code and together performed unit and system testing on all operating routines. At the completion of formal testing, the test environment was released for training in Stage 6.

STAGE 6: Documentation and Training (Change Management)

At the start of the seven-stage methodology, the Imaging Project Team had outlined a documentation and training program that would support development of new processing procedures and user training required to operate the system. In some CBC phases, training on personal computers was required prior to the formal user training activities in Stage 6. In Stage 6, users had to be prepared to absorb documentation and training associated with the imaging system.

- *System Documentation:* As each CBC project phase reached completion, the CBC Imaging Project Team coordinated appropriate documentation with personnel involved with the system. Documentation included 1.) A Systems Administrator Guide to assist CBC system staff in monitoring, performing system backups, handling routine maintenance and addressing simple operational enhancements; and 2.) A User Operations Manual to instruct on how to operate the CBC image system from their perspectives.
- *System Training:* Once documentation was complete, the Project team coordinated appropriate training for system users.
- Training included 1.) Systems Administrator training via a combination of classroom and joint participation efforts to transfer knowledge from the Integrator to CBC. This training focused on design, reengineering and workflow programming. (Advanced training for system administrators is on-going through interaction with the integrator, and formalized Wang/Sigma classes); and 2.) User training which consisted of instructional classes on how to operate the CBC system for management and all claims processing personnel. Integrators performed Train-the-Trainer activities while the CBC ImageProject Team coordinated training rollout. CBC advanced training programs ultimately allowed CBC system users to learn how to build and tailor their own productivity-increasing operating macros.

STAGE 7: Cutover, Acceptance, and Warranty

- *Cutover:* After training, CBC released each phase of image system development into operation. Appropriate support personnel surrounded the users and system in order to address any anomalies in system operation.
- *Acceptance:* The Project team certified system acceptance when completely confident that the system could provide expected operation support to CBC.
- *Warranty:* After acceptance, integrators such as ICG provided warranty on that phase of the system for 90 days. If no integrator was involved, CBC provisioned for warranty type staffing after system acceptance.
- *Maintenance:* Ongoing through all four CBC phases, ICG provided maintenance for the entire CBC imaging system. CBC System Administrators supplemented this effort.

Judgment Factors

The following paragraphs on Judgment Factors (IV) are summarized here for quick review. References are made to the originating Guideline Section to provide cross-reference to more detailed information in the Guideline Questions Sections. In some cases, Judgment Factors have been expanded beyond the topic areas covered by Guideline questions in order to provide a complete answer to the Judgment Factor questions.

1A) Innovation: Innovative use of technology to further strategic objectives

Throughout its four-phased implementation, CBC used a controlled introduction of leading technologies to achieve enhanced productivity. Automated Robots, WAN image distribution, Image Assisted Data Entry (IADE), Priority Routing, Record Data Evaluation (RDE), and Remote Image Workflow are some of the innovative imaging technologies which CBC and other leading imaging installations have implemented. CBC distinguishes itself by introducing technology dramatically to handle three strategic initiatives:

- Transfer of disparate images across separate organizational entities: COMP I Subsidiary required access to claims provided to PBS. To accomplish this, CBC had to:
    - Coordinate with organizations not owned/operated within its direct scope of influence.
    - Design and implement a system to dynamically convert ImagePlus to Omnidesk images.
    - Pioneer use of Sigma's LAN network layer to manage memory overhead.
    - Interface with multiple mainframe systems to pass data and manage image transfer.
- Enterprise access and/or use of imaging throughout CBC: CBC wished to provide access and/or use of imaging throughout its organization. To accomplish this CBC not only implemented a major imaging network, but also provided imaging integration to mainframe terminal users. Specifically CBC:
    - Developed a 300 image users network across four locations
    - Integrated 1,500 mainframe terminal users into the imaging network
- Enhanced workstation automation via user-trained macro scripting and special workstations: Having outfitted CBC with enterprise imaging, CBC wished to ensure that imaging users maintained a constant eye toward researching and implementing new productivity opportunities. To accomplish this, CBC developed a program to train users to develop personal macros, automating individual processing.

1B: Degree of complexity in the underlying business process and IT architecture

The degree of complexity in business process and IT architecture is revealed throughout the discussion of the CBC implementation. Highlights include:

- Complexity in the underlying business process involved:
    - Integrating three separate organizational entities, each under separate control.
    - Centralizing claims systems into one location with operational distribution to five sites.
    - Coordinating business processes and reengineering with PBS.
    - Coordinating business processing and reengineering among three remote sites.

- Maintaining non-disruptive imaging roll-out, avoiding slippage in NMIS rankings
- Fostering user interest, even though 150 positions would be re-deployed
- Maintaining four-phased implementation as structured roll-out toward enterprise imaging.
- Building an internal CBC team to absorb "transfer of knowledge" from integrators.

- Complexity within the IT architecture included:
  - Structuring a high-volume claims processing environment
  - Integrating three mainframe platforms
  - Developing transparent interfaces among mainframes and imaging system
  - Building and coordinating a WAN across 5 sites (3 CBC, 1PBS; 1 COMP. Claim)
  - Printing images from Host terminals
  - Interfacing CBC Image Request Robot (Windows) to the PBS Retrieval Robot (OS/2).
  - Invisible (to user) complete upgrade of technology within last six months.

1C: Creative and successful deployment of advanced workflow and imaging concepts

Advanced workflows and imaging concepts are fundamental to making the CBC one of the more progressive imaging installations in the country. Although the entire Guideline Section refers to various forms of creative and successful deployment of workflow and imaging concepts, the following summary provides some highlights:

- Activities/Use showing creative emphasis on imaging:

Throughout the CBC installations, imaging innovation has been demonstrated by technology use, involvement of the users and orientation of the integrators.

- Used *technology* in innovative fashion (Judgment Factors–Sections 1A; 1B)
- Involved *users* by (Judgment Factors–Section 2B)
  - Providing users with real tools to perform real development themselves
  - Establishing an environment of continuous improvement that has already "paid off"
- Integrators advanced new concepts of
  - Internalization so that CBC could become generally self-sufficient
  - Coordination among integrators for horizontal deployment of imaging across phases of implementation

- Activities demonstrating tangible success of imaging:
  - Installed large high-volume imaging scanning and storage location
  - Maintained user involvement

- Obtained cost/productivity results relayed through MIS rankings
- Evolved through four comprehensive development phases–a unique string of successes
- Maintained cost-justifiability at each phase
- Gradually transferred knowledge from Integrator to CBC

1D: Achievement of Business Process Reengineering and/or continuous improvement

CBC's movement through four phases of development indicates management support of imaging technology benefits largely from reengineering which:

- Automated processing in coding, claims processing and comprehensive claims activities that eliminated many or all of the manual tasks within each function.
- Provided access to a variety of systems for information (multiple mainframes and other image systems) to help coordinate information on the claims processor desktop.
- Allowed for distribution and management of workload among three CBC sites, which gave CBC the ability to dynamically balance workload processing among all its areas.
- Supported ongoing user education and involvement. The success which users feel from their new image-based workflow has fostered a continuous effort to identify opportunities for improvement and use productivity tools to build personal operation macros.

As a result of reengineering, CBC can initiate consolidation of data entry functions from all locations to the central imaging facility. This centralization effort has maximized system uptime, availability, cross training between system administrators, and allowed system administrators to interact also with the users and automate processes.

2A) Implementation: Successful implementation approach

CBC's four-phase implementation over three years involved a detail 7-stage methodology that was launched for each phase of the project. The phases of the methodology included:

- STAGE 1: Project Initiation and Management (Project Team/Change Management)
- STAGE 2: Project Analysis (Reengineering)
- STAGE 3: Project Plan and Requirement Development
- STAGE 4: Planning and Design
- STAGE 5: Installation, Development and Testing
- STAGE 6: Documentation and Training (Change Management)
- STAGE 7: Cutover, Acceptance and Warranty

The result of this implementation methodology was a series of four successful installations and true transfer of knowledge to the CBC internal development team.

2B: Extent of change management process

CBC aligned its change management program with each phase of the implementation. The program involved two initiatives:

- User Education:
  - Image/Workflow Project Education: User involvement in all stages of each phase
  - Technology Education: Updating users on emerging technologies
  - Post Implementation Review: Reviewing and reexamining the system for opportunities
  - System Documentation and Training:
  - Documentation for users and System Administrators for all system in all phases
  - Training programs for users and System Administrators for all system in all phases

The result of this advanced level of Change Management is that CBC users have embraced the system and feel not only ownership for the system, but also pride in what the system has been able to accomplish, with their support, for the final CBC system.

2C: Level of overall system complexity

System scope, processing capacity, distribution, integration, technology and orientation to upgradeability continue to contribute to the overall level of complexity for the final CBC system:

System Scope:	300 direct image system users 1500 host-integrated users
System Processing Capacity:	4 high-speed scanners 4 high-capacity jukeboxes 8 servers 12 printers 8.5 million images per year
System Distribution:	5 sites 3 organizations Multiple high-speed lines
System Integration:	2 mainframe systems 2 imaging systems
System Technology:	Image conversion/transfer Automated Robots Color monitors (one of 1st in Industry 1993) Memory management using Wang LAN network Workstation automation via macro scripting Printing from the (1500) 3270 terminals
System upgradeability:	Strategic Hardware Selection: followed vendors growth path while not obsoleting last year's purchase Jukebox (4)— HP 100T 60 MB drives Upgraded to 200T 1.3GB drives 600 platters still valid in new drives Holds 1.5 years on line Scanners (4)— Kodak 900. Upgraded to 923.

Monitors and Windows Accelerator Cards:
Pilot used 19" IBM 8508 with Image Adapter A (approx. 2,000). Phase II tested new Artist Graphic Card with Hitachi Monitor, 20" color (approx. $2,000). For the same cost the monitor was color, bigger and had faster image display times than most market-available setups.

2D: Level of integration with other advanced technologies

The CBC Claims Processing System is unique because it integrates three major information processing environments:

- The CBC mainframe environment including CBC AIRS' claims processing system
- The PBS mainframe environment including PBS claims processing system
- The PBS ImagePlus image processing system

3A.) Impact: Extent and impact of demonstrated productivity improvements

Specific productivity improvements realized are attributable to an increase of more than 25 percent in productivity as a result of the imaging system. Improvements include:

- Eliminating all existing backlog
- Reducing claims processing time by minimum of seven days
- Reducing access to PBS claims from 3 days to 30 seconds
- Enabling Customer Service Representatives to answer a question on the initial call
- Reporting statistics on performance to restructure work profiles and optimize effectiveness
- Supplying imaging tools which allow the users to build their own productivity-enhancing system features

3B: Significance of cost savings

Specific cost savings that have been derived from the implementation of the claims processing systems aggregates to over $3.5 million over a five-year period and are expected to yield an additional $1.1–$1.9 million each year. These savings were generated by:

- Reducing claims processing staff and support personnel 80-110 employees
- Reducing the document preparation, batching and filing staff by 20-40 full-time employees
- Eliminating microfilm/microfiche servicing costs; calculator and calculating costs

3C: Level of increased revenues, product enhancements, customer service or quality improvements

CBC has generated opportunities for new products, more efficient customer service and improved quality through the introduction of the Claims Processing Imaging System:

- Improved Revenue/Product Opportunities by:

  - Freeing time for managers to meet on new products/processing strategies
  - Evolving workflows into new forms of service offerings
- Improved Customer Service By
  - Processing claims seven days faster than before implementing imaging system
  - Reducing need for call-backs
  - Providing quick access to customer information
  - Providing quick communications (letter/fax) to customers
- Improved Claims Processing Quality by:
  - Introducing data/management tracking reports
  - Removing manual intervention (and possibility for error) through workflow
  - Performing edit checks and lookups associated with Image Assisted Data Entry (IADE)

3D: Proven strategic importance to the organization's mission

The National Management Information System (NMIS) creates standards for excellence in service quality and costs for Blue Cross and Blue Shield Plans in order to ensure consistency and quality within the Blue Cross and Blue Shield Association. Each Blue Cross and Blue Shield Plan must meet or exceed NMIS standards and each year Plans are ranked by their performance against the NMIS standards. Competition is keen. Both service quality and costs must be managed equally— with little option to raise cost or lower service levels.

By mid 1995, CBC had experienced more than a 25 percent increase in operating performance. Over 100 employees had been freed up by imaging so they could be allocated to other operations, while eliminating the need for night shift processing. In addition, the amount of time required to process a claim had diminished by more than seven days.

3E: Degree to which the system enabled a cultural change within the organization

The CBC approach promotes an environment for continual improvement and reengineering, and by receiving a "transfer of knowledge" from integrators to building internal support and independence:

- Foster environment for continual improvement:
  - Ongoing imaging education classes
  - User leadership and involvement in any systems development
  - User training on productivity tools to build personal operating macros
  - Team meetings to discuss improvements to imaging systems
- Transfer of knowledge from Integrator to CBC–for internal capability and independence by Phase III:
  - Phase I: ICG and Wang

- Phase II: ICG
- Phase III: CBC with IBM and ICG specialized consulting
- Phase IV: CBC with ICG specialized consulting

3F: Impact of system on competitive positioning in marketplace

CBC has become more competitive in the marketplace. This move resulted primarily from the following:

- Improved productivity by 25 percent
- Reduce costs by $3.5 million and $1.1–$1.9 million annually
- Improved ability to launch new products by freeing available resources to meet improved customer responsiveness by reducing callbacks and providing quick access to information.

Consolidated Edison of New York
New York, New York

North America Excellence Award: Workflow, Silver

Executive Summary

The Consolidated Edison Automated Workflow System (hereinafter referred to as AWF) was initially implemented in 1992. Since the system was first installed Con Edison has added significant features which support their customer service objectives. The AWF system is used by approximately 600 Customer Service Representatives (CSRs). When they first installed the system these employees worked on three types of paper:

- Customer correspondence (one million letters per year)
- Faxes from government organizations (100,000 faxes per year)
- Mainframe documents which were sent to printers (several million "citations" per year). These mainframe documents are citations produced each evening in the batch process, which highlight potential problems with accounts

These documents, in their electronic form, are now processed through the AWF system throughout the day. Utilizing Wang Laboratories Inc.'s OPEN/workplusTM RouteBuilderTM software, Con Edison is able to direct the flow of each of these electronic documents from their entry into the system through their completion.

1.) Describe the system application. What the system is used for, who are the users and what does the job entail?

A typical document flow would include the following steps:

- Work would be loaded into the system via one of three methods. Customer correspondence would be scanned in, and fax documents would automatically be entered as electronic data. Mainframe documents are sent directly to the AWF server, using a "loader" program developed by Wang.
- Electronic documents are routed to predefined queues based on several criteria; type of work, dollar amount, age, etc. The routes are changed by non-Information System personnel to reflect the changing needs of the organization.
- CSRs are assigned to work these electronic documents, either as fill-in work or for the entire workday, depending on needs.
- Work is sent to succeeding steps in the route, until they are either sent for deletion, sent to file, or sent to a Quality Review queue for analysis by supervision.

This use of the AWF system differs from the customary use of workflow systems, both technically and functionally. As a result, Con Edison has new ways of achieving organization goals. Several examples follow:

- The mainframe documents are loaded into the servers for routing to CSRs. This method bypasses some of the actions needed by companies who use COLD (Computer Output to Laser Disk).
- The work is no longer assigned only to "back office" personnel. Work is routed through the Wide Area Network (WAN), as fill-in work to employees in the Call Center and walk-in Service Center.
- Several employees work at home, using ISDN telephone links to the AWF servers. Thus, employees at home can do the same work as they did in the office, with the same or better productivity and quality.
- "automated quality review" as an integral part of the routes
- Off-the-shelf ad-hoc reporting software is used with Wang software to provide improved management reports.

Thus, CSRs are able to respond easily to customer correspondence, answer faxes, and process mainframe citations.

2. What are the key motivators behind the system?

Originally, the key motivators were cost reductions. In 1990, Con Edison had 240 clerical employees supporting their Customer Operations organization. An industrial engineering study confirmed that almost 80 percent of the clerical functions were related to "paper-pushing" tasks that provided no real value to customers.

Con Edison also surveyed 400 of employees, including CSRs, their Managers and some Executives. The survey results emphasized that there was "too much paper", and they were losing efficiency due to this paper excess.

A six-month pilot test conducted in one branch office determined that they could realize significant clerical savings with an AWF system. Four years later they have fewer than 40 clerical employees, a significant reduction in personnel attributable to many factors, including AWF.

Yet, even without this extremely favorable payback (less than one year) they would have installed the AWF system, reports Ed Glister, Senior Project Manager. The organizational flexibility the system provides is greater than could be realized any other way.

Previously two organizational factors were barriers to improving work processing:

- The system was organized into three functional groups: back office, Call Centers, and Service Centers. These organizations never shared work in the past, due to their specialization.
- Employees were working from physically remote sites throughout the New York metropolitan area.

The AWF system overcame both these barriers. The organization was changed so that employees worked on different tasks during the day. As an example, a CSR in the Manhattan Call Center may work mainframe citations at one time of the day, and answer phone calls at other times. If a site is backlogged, some remote site can process their work, by accessing the work over the WAN. CSRs can now access work from their home, regardless of where the work may exist within the AWF system.

"For us, these organizational benefits, and other intangibles were more important than the significant tangible cost reductions realized," says Ed Glister.

3. Please describe the current system configuration (number and type of software, servers, scanners, printers and storage devices, etc.).

Con Edison has nine primary server sites. There is one server in each of six headquarter-offices (the five boroughs of New York City and Westchester), which service most of their 600 users. Additionally, there are three dedicated servers for special tasks.

The servers are Compaq XL dual-486/50 mhz processors, with one gigabyte of memory and RAID V. The CSRs use Dell 486/33 Model ME PCs and have 17" NEC 5FG displays.

Data storage is accomplished on the server for the first 90 days then sent to HP jukeboxes when storage is required. The six jukeboxes range from 20GB to 60GB in size. There is one or more Fujitsu scanner at each site. The software, when purchased, was Sigma's Omnidesk (now acquired by Wang Laboratories Inc.). This utilizes OS/2 and Database Manager. There is one site using Wang's OPEN/workflow 3.0 software that utilizes NT with SQL Server as the database.

4. How is this system integrated with the Company's other information processing systems?

The integration of the AWF system with other Company systems occurs on various levels. This project was installed as part of a larger project called "New CSS," the new *Customer Service System.*

It was installed as part of a project that included PCs for 1,500 employees, a WAN at T-1 speeds for 31 sites, and five major client-server software applications, which included:

- a graphical user interface
- telephony (voice-data integration)
- ad hoc reporting
- automated cash processing
- automated workflow

These systems are described in a videotape, *New CSS 93/94*, which was produced to inform CSRs and management of the applications being installed. It provided a quick way to understand the scope of the New CSS project and specifically the AWF project.

When the project first started each of these applications was included based on its own benefits. With the ensuing years it was found that the integration of these applications yielded greater benefits.

The first integration action related to the linkage of the AWF with the mainframe system. When a CSR starts working on an item in AWF, they press the "Control-A" key and initiate a Visual Basic application that accesses the proper mainframe account, and the proper mainframe screen for the document. The next step of integration was extending the use of the ad hoc "point-and-click" reporting tools from the mainframe to the AWF system. Thus, *Quest* and *Forest and Trees* are used in AWF, enabling analysts to use only one query tool to meet all their needs.

The next aspect of integration was the addition of the system with other applications, so that CSR could use any system at any time. This provided the work flexibility noted above.

Use of "home office" capabilities of the Wang product enabled extension of the complete office environment to the CSR's residence.

Con Edison is now about to use the product in yet another unique way—to integrate it with the Call Center Automation Call Distribution (ACD) software. In this manner the CSR who is not busy will be automatically assigned an item or work from the AWF system, using DDE links.

5. Describe how the Company has been impacted by this system. Be as specific as possible:

a. What cost savings or increased revenues have been realized since the system was first installed?

b. What productivity improvements have been realized?

c. How has the business workflow been affected (compared to before system implementation)?

As described above, Con Edison realized tangible savings related to clerical reductions. These are estimated to be worth $2 million per year.

"While the benefits of organizational flexibility are difficult to quantify, we feel they are more significant than the clerical savings. Each Manager has the ability to change the way work flows through the organization, by using the RouteBuilder tool. This means new methods of processing work can be tested and installed overnight," says Ed Glister.

"And, when backlogs arise in one site and workers are idle at another, we can immediately transfer work over the WAN, a benefit not achievable in the paper world."

There are other intangible benefits as well. Employees are more satisfied using the new system. It responds to their complaints about paper, and makes it easier to service the customer quickly and with better quality. Also, management is more effective, since the automated quality review feature guarantees that they are reviewing work in accordance with their own guidelines. Further, the AWF system provides on-demand reports regarding cases in inventory and detailed audit trails to satisfy security requirements.

"And let's not underestimate the value of finding work when it is needed, overcoming a problem that existed with the filing of paper documents."

Finally, the ability to balance different kinds of work on a CSR's desktop is also valuable. Employees get tired of working on one task. With the integration of the AWF system and other new automated systems, they are able to vary workloads to employees, keeping them interested and productive.

From a workflow perspective, the way work is processed is totally different than it had been prior to the system. No longer bound by organizational charts or geographic limitations, each manager changes the workflow routes when needed, and they do so without the assistance of an I.S. programmer.

Not only is this a *potential* benefit, it is a real benefit. This "reengineering on the fly" is easy to introduce, is clearly outlined in the new route maps, and can be changed anew if required. Initial routes had only 30 processing rules, now they have over 800 rules. The value of this tool is evident each time routes are restructured.

6. Describe the implementation process and methodology, the project team and any change in management and business process re-engineering issues addressed.

The implementation follows:

- After determining what users wanted, Con Edison developed a six-page list of functional and technical requirements for the AWF system.
- The AWF system was introduced in a pilot test in the Mt. Vernon branch office. (Mt. Vernon management had volunteered to "champion" the pilot test.)
- At the end of the six-month pilot test CSRs and Managers from five other organizations were invited to use the system for one week. This ensured that they didn't have to make a decision based on a demo but, rather, could use the system themselves to get to truly understand it.
- Prior to installing the software an educational campaign, consisting of information videotapes, demos and newsletters, was initiated. This way CSRs learned about the system with a consistent message.
- High-speed lines and PCs were installed prior to installing AWF software at each site. This provided immediate benefits to the CSRs, 90 percent of whom had never used a PC before the New CSS Project.
- The electronic forms were designed the same as the paper forms used previously. This reduced the learning curve for employees.
- Software was available to the best users first. This reduced novices' fear and also ensured the best feedback initially.
- The training was developed by professional staff, and administered by respected CSR supervisors. A two-day training class included computer-based training and use of the new system. On-the-job training was available from staff support and others who had completed the course.

The AWF system and the other four systems identified above were installed during an 18-month period. This schedule enabled the CSRs to gain confidence in the new system and to realize the benefits as quickly as possible.

Once the implementation phase was completed, the system went through new phases of development. Some of these phases were occasioned by changes in organization. During implementation, Call Centers dropped from 23 to six and several functions were centralized, which resulted in the use of the AWF for these newly constituted organizations. In addition to these large-scale changes, users made other changes, as they became more familiar with the benefits of the routing tool.

The installation team consisted of two employees from the Central Customer Operations staff, two LAN support specialists from IS and others from Training Services and the Operating Organizations. Consultants were not used during installation.

In summary, the AWF system has provided Con Edison with a platform to meet Company goals and serves as a platform for future integration and enhancements.

The following information has been extracted from the December/January 1995 issue of Review "Training and change: the key to reengineering":

Number of Customers Served	5,000,000 at 19 Site Offices
Number of Phone Calls Annually	6,000,000
Number of Documents Received Annually	• 4,000,000 managed by 200 clerks • 80 percent of clerk's day spent on paper process • Paper physically moved to various site offices for work load balancing during peaks
Reengineering Efforts	Begun December 1992 • Team of 15 Employees: Manager, Union Personnel, and Information Systems (IS) • Week long Joint Applications Design (JAD) session facilitated by IBM
	Goals: • Reduce number of Call Centers from 23 to six • Answer Calls 24 Hours per Day (vs. 8:30 a.m. to 5 p.m. Mon.-Fri.) • Reduce time to answer call from 40 seconds to five seconds • Reduce costs by 25 percent in three years • Increase workload by 20 percent • Conducted 400 employee interviews after JAD reengineering session. Over 90 percent of employees had never used a PC before and were concerned the company would not train them properly

Customer Surveys Included	Employee involvement: • Must become stakeholders in design/implementation "Champions" concept introduced by department to lead project and champion results • Brooklyn office request to champion telephone Center • Originally planned pilot with 10 users • Opted for more aggressive approach installing a LAN, 100 PCs and new software in 90 days • Westchester office request to champion Customer Service process • Gradual introduction of workflow and imaging technology over six months to test alternative ways of doing business • Implemented concept of "user self-sufficiency" • Software enabled them to create/modify own reports • Software enabled end-user organization to define/modify workflows or create ad hoc reports without IS involvement
Corporate Workflow & Imaging Rollout	Implementation Team • Central Customer Operations Staff • LAN Support Specialists • Training Services and Operating Organization. New technology reduced system response time from 3-6 seconds for data to appear on screen to a "split second."
External Audience Test	• Presented to Public Service Commission • Each gas and electric utility is subject to review by PSC • PSC needed to authorize expenditure for project hardware and software • Test was testimonial of the reengineered processes and the capabilities of new technologies
Industry Coverage	10 articles in publications during 1995 including • Business Week • ComputerWorld • Software Magazine • January 1995 included in Giga Review May 1996 Microsoft Solution of the Month
The System 1996	• Users at nine primary server sites • One server at each of six headquarters offices (five boroughs of NY & Westchester) • Three dedicated servers for special tasks • Six jukeboxes ranging from 20GB to 60GB • One or more Fujitsu scanners at each site • Wang's OPEN/workflow and OPEN/image on OS/2 and

	Database Manager • One is using Wang's OPEN/workflow 3.0 on NT and SQL Server
Results	• $2,000,000 in annual savings related to clerical reductions • Customer Call Centers reduced to six • Ability to find work when needed • Computer-literate organization

First Albany Corporation, Albany, New York

North America Excellence Award: Imaging, Finalist

Executive Summary

Driving Growth with Workflow-Enabled Business Process Reengineering

Business process reengineering (BPR) often is viewed, especially by workers, as a euphemism for corporate downsizing. However, at a forward-looking organization such as First Albany Corporation in Albany, New York, reengineering fulfills its true promise—as a strategy for rapid growth and complete customer satisfaction in a business environment of intense competition and constant change.

To implement reengineering, this investment bank has combined a highly structured BPR methodology with essential business process automation (BPA) tools, including client/server workflow software, imaging technology, and a computer output to laser disk (COLD) solution. In one key area-customer accounts processing and records retention-the results to date have been remarkable.

Large amounts of paper output and physical storage for hundreds of thousands of account documents have been eliminated. Procedures have been streamlined, reducing processing errors from 50 percent to five percent while collapsing process transfer times. First Albany has expanded its productive (sales-related) workforce by approximately 15 percent during the past two years while maintaining the same size support-related workforce. All together, these results are helping First Albany achieve its ultimate goal of providing "second-to-none" service leading to complete customer satisfaction.

The Challenges

First Albany Corporation, the main subsidiary of First Albany Companies Inc., is an investment banking and securities brokerage firm with 650 employees and approximately $120 million in annual revenues. The organization provides institutional securities brokerage services as well as retail brokerage through 35 branch sales offices concentrated primarily in the northeastern U.S.

As a securities dealer, First Albany Corporation is fully regulated by the Securities and Exchange Commission, the New York Stock Exchange, the National Association of Securities Dealers, and other agencies. As a result, First Albany must track thoroughly all transactions and retain voluminous documentation to fulfill regulatory and reporting requirements.

Until recently, First Albany's regulatory responsibilities translated into paper— vast amounts of it— as well as microfiche records that needed to be filed for years, often indefinitely. One of the most prolific producers of paper was the customer accounts area. Both regulatory and customer service requirements dictate use and retention of up to 50 different forms.

"The requirement is that these forms must be retained for as long as the account remains with the firm," explains Ed Brondo, chief administration officer for First Albany Corporation. By the end of 1993, he continues, "We had about 500,000 documents supporting a base of approximately 120,000 customers, and we were opening 50 to 100 new accounts a day."

At that time, in 1993, the information technology infrastructure was minimal at First Albany. There were very few personal computers (PCs) and no networks. Securities transactions were processed via a service bureau, which owned the mainframe computers and provided First Albany with access through "dumb" terminals.

As Brondo explains, "We would have some downloads of hard copy for our database, and that was the state-of-the-art. We needed something for budgeting, planning, and in-depth financial analysis. Use of the service bureau for control and analysis simply wasn't working for us."

Brondo was hired in June 1993, and charged with changing the situation through process reengineering and automation. By December 1993, First Albany Corporation had signed a contract with Computron Software, Inc., to implement financial software, a COLD solution, and workflow. During the spring of 1994, the computer hardware and software foundation was laid: local area network (LAN) and IBM RS/6000 server installation; PC installation and training; selection of an NKK optical "jukebox" for high-volume information storage; and conversion of service bureau information to flow into the Computron system.

By May of 1994, with both the information technology infrastructure and a business support analysis group in place, First Albany was prepared to launch its various process reengineering and automation projects. What follows is a detailed description of the efforts that ensued.

Workflow-Enabled Customer Accounts-Reengineering Implementation

On October 1, 1994, First Albany "went live" with its first ever in-house financial software system-accounts payable and general ledger modules from Computron Software. However, "that wasn't enough to keep us busy," Brondo says. "At the same time, one of the big things we wanted to accomplish was to eliminate paper."

As noted above, the firm was retaining approximately 500,000 documents, and adding between 50 and 100 new accounts daily to the pile of paper.

"Trying to locate those documents could be a nightmare," Brondo notes. "We wanted to eliminate several thousand square feet of file space and replace it with an optical 'jukebox' that takes up a couple of square feet of floor space."

By September of 1994, First Albany had installed Computron's Workflow product and was prepared to scan documents into the imaging system.

"In September, we decided that all new accounts would be scanned into the optical system. Then we started a project to work backwards scanning the records of current accounts.

"We now have a workflow system in the neighborhood of 300,000 documents, and we continue to scan all of the new documents opened every day," Brondo continues. "All of these are readily available, all are indexed and easy to find. This technology also satisfies the regulatory bodies-it has to be WORM (write once, read many) technology, and it has to have an off-site backup."

Beyond Paper Savings

Paper and space savings, however, represented only the beginning of what First Albany wanted to achieve with imaging and workflow. The company's business support analysis team, formed in the spring of 1994, had determined by fall, 1994, that significant gains could be achieved by reengineering and further automating the customer accounts documentation process. This process generally originates in First Albany's branch offices, then continues after documents are sent to the central offices in Albany, New York, for processing.

"At the time, the error rate in the document completion process was about 50 percent. Therefore, we were constantly sending documents back and forth between people and offices," says Brondo, "The other thing we found was that we had about 15 steps and 13 different 'hand-offs' from initially filling out forms through approvals. We decided that didn't make a whole lot of sense.

"So we put together a design team composed of people from operations, sales, and support. They came up with a process design and determined that we could ultimately reduce the number of steps from fifteen to five—though in the interim, the number of steps would be reduced to eight."

First Albany also realized that the process would be streamlined simply by scanning documents into the workflow system at the branch offices. Though scanners are gradually being installed in branch offices, the simplest way to capture document images was with the fax machine.

"Faxes are captured as images automatically by our workflow system, then identified as to the branch, then transmitted to our new account department electronically. The documents are put in a work queue. New account personnel can call up images, review the documents for completeness and accuracy, index them, and store them in the optical system. That entire flow now has been automated, eliminating all the paper for customer accounts going from branch offices to the main office."

The combination of workflow-enabled automation and process reengineering has not only eliminated paper and reduced process steps, but also slashed the error rate from 50 percent to an average of five percent and even one percent at many branches.

Furthermore, "In addition to maintaining the basic account documentation, we have to confirm every trade and transaction a client requests," Brondo says. "We send out a confirmation, and copies of those are stored optically and linked to the other account documents. So we're winding up with a rather extensive picture of the particular client, which is rapidly accessible through jukebox storage."

Further Reductions with COLD

During this same time, First Albany installed a client/server COLD system (Computron's COOL, or Computer Output On-Line) to further reduce paper printing and storage.

"To give you a sense of the level of production, we used to print on a regular, daily basis six full boxes of greenbar paper reports— over 700 reports daily from our service bureau directly relating to our reporting requirements," Brondo explains. "We don't print them any longer— we store them in COOL and people access them from their desktops. As a result, printing has been reduced from six boxes to two boxes."

These COLD documents share storage space with workflow documents on the optical jukebox. Also, many of these reports previously had to be stored on microfiche, which now has been eliminated.

"We're saving on computer time, printer time, the bursting and distribution process, the cost of paper, and the cost of microfiche," Brondo says. "And we're saving the cost of people's time researching through microfiche and paper files. That used to take hours, and now people can retrieve answers in seconds. So our estimate is that we are saving about $100,000 a year with COOL."

Workflow—An Enterprise-Wide Impact

The powerful impact of First Albany's customer accounts workflow application is being felt across the entire organization.

"This particular application most directly affects the retail segment of our business," Brondo explains. "Out of 650 employees, something like 400 are directly related in some way or another to retail-so workflow directly impacts 400-plus people."

Currently, direct access to the workflow system and imaged documents is available in the central office. However all branch offices (including 12 that are not yet equipped with a PC LAN, but still have at least one PC) are connected to the home office through high-speed ISDN telephone lines. These lines allow hourly database replication of the company's enterprise-wide Lotus Notes application. Computron Workflow is linked to Lotus Notes, which allows rapid delivery of imaged documents (via fax or electronic mail) to any location in the First Albany organization. And this is a stopgap measure.

"Once we've installed networks and PCs in every branch, they will be connected to the central system in Albany by a frame relay network," Brondo says.

"Then the workflow-based customer accounts system will directly impact almost everyone in the company."

Since 1994, workflow already has made it possible for First Albany to add some 100 "producing" or sales-related employees to the firm while staffing levels in support areas have remained flat.

"We've had very sizable growth of personnel on our revenue side, but haven't had to add people to support them."

Moreover, the revenue-generating members of the workforce have become more productive as well.

"If our document error rate with new accounts was 50 percent, there was a sizable amount of time that used to be wasted," Brondo says. "Now since we've reduced errors to five percent, the 100 or so sales assistants who are involved in opening new accounts each day now are freed up to directly support our client base.

"So I have to believe that we have more satisfied clients, because there are more people who can service the clients quickly without being hung up in clerical nightmares."

The Future...

Next on First Albany's reengineering agenda is to workflow-enable its accounts payable department. As with customer accounts, Brondo anticipates a powerful impact despite the small size of the accounts payable department, which has two full-time employees and about five others who work there part-time.

"Beyond those people, we have some 200 managers who spend time approving transactions or otherwise getting involved in the accounts payable process," Brondo explains. "So it's currently cumbersome. It's not only the paper wasted. Our mailroom is sending paper all over the firm. The transmission time is long, whereas an image gets there in a few seconds. Now it's a real trick to determine where approvals are, who needs to sign-off next, and so on. With workflow, we'll know exactly where everything is, and it will speed up the whole process and make it more efficient.

"Managers who waste time handling paper can do the approvals right on-line. The incremental savings will add up to more time people can spend with their clients, and in general, more time they can be productive. We really want to free up time for the managers. That's the main goal."

The Methodology

In the spring of 1994, First Albany formed a six-member business support analysis team. The team's first critical task was to document, through interviews and other research, all of the individual procedures that comprise the company's operations.

"They spent about six months in that documentation process," Brondo explains. "What they came out with was a book of about 600 pages of flow charts and related narrative describing procedures."

From there, the team began analyzing specific functions that integrate various procedures into a process flow involving a number of individuals. For example, in the new account process, the analysis team determined what it took to open an account, what the client does, what the salesperson and sales assistant do, the mailroom clerk, everyone who touches that process.

"Then we put together a design team that included representatives from each area impacted by the process flow, perhaps 10 people," says Brondo. "Their purpose was, within the framework of six well-defined design criteria, to figure out the best way to perform a process."

Those criteria are as follows:

1. Any process has to be done to provide quality customer service.
2. Speed in process is equal to quality. "If you're a customer who calls on the phone and somebody has the correct answer immediately, then you feel comfortable about First Albany," says Brondo.
3. In order to have speed and quality, you have to have simplicity.
4. Simply offer the right choices. "In many cases, there are multiple ways of doing something, but all are not efficient or good for the customer. We define what is good and offer only those choices."
5. Done once and done. "We want people in jobs who are trained, who understand what is expected, and who do things right the first time. Correcting errors takes much more time than doing it right the first time."
6. Simply try to achieve a paperless environment.

Once the process design team, working in concert with the analysis team, develops a new design, the team makes a formal presentation to a steering committee comprised of senior managers representative of impacted areas. The three teams then hammer out agreements on the process redesign, and the business support analysis team manages and implements the changed process. In areas where workflow technology is key to implementing the reengineered processes—such as the reengineered customer accounts documentation process—the analysis team works closely with Computron consultants to "convert our thinking into the actual workflow."

"We've completed this procedure with approximately 10 areas of the firm besides new customer accounts, and we have another 30 or so to work on," says Brondo. "A lot of people talk about reengineering, but they don't have senior management support or a formalized process to make sure it happens. We have the design criteria, the basic documentation, the formal process and top-level support for the effort."

Gak Nederland b.v.
Amsterdam, The Netherlands

European Excellence Awards: Workflow, Gold

Executive Summary:

Like many institutions across Europe, Gak Nederland b.v., the largest Dutch Social Security provider, is spearheading the drive for efficiency throughout its organization in order to control costs and spending and provide better and more efficient services to its public.

One department within Gak Nederland b.v. is leading the way: the Employers Insurance Administration Department, in conjunction with Gak's Document ConversionCenter Department, recently succeeded in reengineering the process of registration, premium calculation and filehandling. This department serves more than 280,000 Dutch employers for the insurance of their seven million employees/beneficiaries, accounting for almost 65 percent of the Dutch workforce. Gak Nederland b.v., distributes $11 billion to beneficiaries per annum.

Previously, the Department handled over eight million filed documents manually. "We had major problems due to the amount of paperwork involved, information was not easily accessible, files often got lost and misplaced, thus premium processing was delayed. We had no adequate means to manage those complex processes," says Manager Johan Heetwinkel.

In order to facilitate such business critical processes, various logistical issues needed to be reengineered. With the introduction of a data logistics concept, ASZ, an IT company of the Gak Group, provided its Customer with a rich toolset for Work Management. "We defined four layers; strategic, tactical and operational management and the operational process itself." said Jaap van Zetten, Marketing Manager for ASZ.

ASZ implemented a tactical steering layer with BAAN (Triton) industrial logistics software (installed at Boeing as their logistics application), and Wang Laboratories, Inc. (Wang)'s OPEN/workflow (IBM/AIX), OPEN/image (Sun Solaris), PC DOCS—DOCS OPEN (Windows NT Server), Netstor (Hierarchical Storage Management Software on Windows NT Server), and Anacomp COM application as operational instruments. ASZ integrated those components with existing legacy (approximately 150 VAX-based systems as a result of right sizing from ICL mainframes) systems and thus implemented a full-scale and operational solution.

Currently the department electronically processes some 280,000 files per year, each approximately five to seven pages thick and growing at a rate of 25,000 files per year, with more than eight million image documents on line today. As a result of implementing a redefined logistics in an administrative process and by integrating Wang's work management software, all relevant information is available immediately to our Customer service staff, thus improving their responsiveness and productivity. Processing time has been reduced from one month to two weeks and we have also saved two floors of office space (more than 1,600 square meters, representing approximately 500,000 guilders now occupied by office personnel), eliminating paper document storage completely.

Gak's staff and customer service representatives use Wang's OPEN/image software to scan approximately 900 documents per day including correspondence, income statements etc. PC DOCS— DOCS OPEN, manages the process of indexing, cataloguing, processing, managing and retrieving these incoming documents as well as Gak's eight million on-line files.

The staff can locate a document anywhere on the system at any time improving responsiveness and productivity. They can retrieve files and images via PC DOCS— DOCS OPEN based on certain criteria and view the images in OPEN/image software. Despite the high volumes involved, the system gives users immediate access to files.

Wang's OPEN/workflow software is used to route the scanned information, through the system electronically to different areas of the department for further handling and investigation. This application also assigns tasks to the Gak staff members, alerting them automatically of any outstanding actions which need to be taken.

"We chose Wang because of its work management architecture and ability to easily integrate with open and standard solutions like PC DOCS — DOCS Open. This seamless approach to process information has brought our client significant business benefit and cost savings of up to 25 percent. We expect to extend this approach to other clients in the future", says van Zetten.

Users access the system by an easy to use Microsoft Windows front end. The system was integrated into a multiple platform environment, including Sun Solaris, IBM AIX and NT Servers by ASZ. Wang's OPEN/image runs on Sun Solaris and OPEN/workflow is running on an IBM RS6000 Servers, while PC DOCS-DOCS OPEN is running on the Windows NT server. These are linked together via Ethernet and a TCP/IP protocol. Documents are scanned by four dual-page Bell&Howell scanners and archived using Netstor HSM Windows NT Server software onto HP jukeboxes with 800gb of storage space.

Section 1. Describe the system application. What the system is used for, who are the users and what the job entails. How often or how many hours is the system in use on a daily basis.

1.A Company Background: Gak Nederland b.v.

Gak Nederland b.v. is the largest social security provider in the Netherlands. Gak Nederland b.v. provides sickness and unemployment insurance serving 280,000 Dutch employers and institutions for the insurance of seven million employees and beneficiaries, accounting for almost 65 percent of the Dutch workforce. Gak distributes $11 billion to beneficiaries per annum. The organization covers over 30 district offices throughout the country and has 13,500 employees. Gak Nederland b.v. is a member of the Gak Group. Gak Nederland is spearheading the drive for efficiency throughout the organization in order to control costs and spending and provide better services to its public.

Two Departments are leading the way:

- Employers Insurance Administration (EIA) Department, which registers the Employers and prepares premium calculations and collection.
- Document Conversion Center (DCC) Department, which handles high volume conversions of paper based, data and microform documents to other media types.

ASZ automatisering sociale zekerheid b.v.

ASZ, Automatisering sociale zekerheid b.v., is an IT Company within the Gak Group, servicing Gak Nederland b.v. and other clients with its staff of 700 people and a variety of IT services such as data center services, project management, software development, business process redesign, document information systems, imaging and workflow and data logistics.

Project participants:

- Gak Nederland b.v.
- ASZ Automatisering sociale zekerheid b.v.
- Wang Nederland bv
- Anacomp B.V.

1.B Document Conversion Center-DCC

Back-office bulk conversion to a mix of storage media

Gak's Document Conversion Center (DCC) uses a mix of storage media. "There isn't a single medium that can meet all our retrieval and storage needs, now and in the future, so we've opted for an optimum combination of storage media: paper, 16 mm roll-film, microfiche jackets, COM (image) fiches, optical disks, rewritable MOs, 8mm tapes and CD-R," says Pieter Minten, responsible for document information supply.

In order to structure various media types, the latter part of the existing files and other information carriers needed to be structured in a usable form. In a joint cooperation between Anacomp B.V. and Gak's Document Conversion Center, various media types were structured in a usable mix such as:

- Microfiche for longer-term storage as back-up medium
- Jackets and 16mm film (CAR)
- Paper, since the fact that not all organizations use the same/compatible resource 8 mm tapes as intermediate medium after bulk scanning to feed the COM document and information systems.
- Rewritable MO disks to store imagedocuments
- Roll-films for feeding digital workstation
- CD-R for recording bulk mainframe correspondence
- In addition, separate OMR and OCR projects have been set up
- Scanning of micofilmed documents to TIFF-files

The incoming information, from wherever and in whatever form, needed to be converted to an industry standard electronic format TIFF (Tagged Image File Format). This forms the base of Wang's document/imaging information system and was processed via various systems. Paper documents in a variety of formats are currently scanned at a rate of approx. 500,000 a month. Historical microfilmed documents are scanned at a rate of 250,000 a month.

Paper scanning is done via four Bell&Howell bulk scanners, creating the TIFF images to be indexed in PC DOCS-DOCS OPEN and stored in Wang's OPEN/image system. Another not insignificant aspect of the route chosen by Gak was the manageability of costs. Converting a single paper document need not be expensive, but millions of documents to be used in a document management system can be considerably costly. This leads automatically to the mix of media to be used to date.

Anacomp's XFP 2000 COM system appeared the ideal system to create usable micrographic information carriers. The bulk of the millions and millions of paper based files were processed by Anacomp's scanning and conversion technology and served the ideal starting point of Wang's integrated imaging and document management system servers for in depth indexing and archiving.

The conversion of all those types of dissimilar media types is a complex matter, which as an individual complementary project was well coordinated and served by Gak's DCC, assisted by Anacomp. Conversion is the crux of the matter. The document conversion allows for Gak to deliver information to their external customers in the form they can use. Internally it allows departments to have access to the information as they build their own infrastructure for the work management solution, and provides a secure backup of the information that is maintained for decades by Gak.

Document Conversion Center (DCC) Department Process

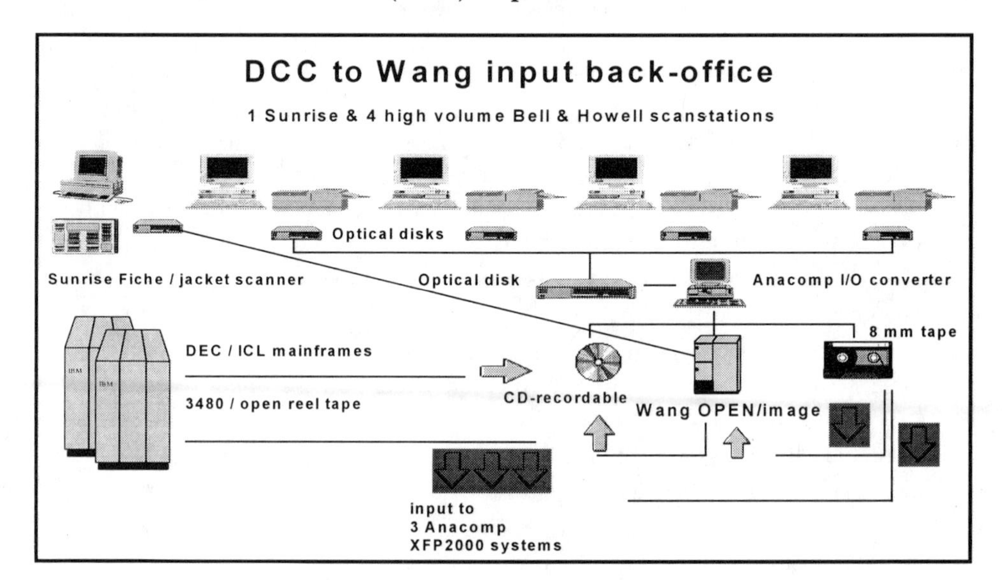

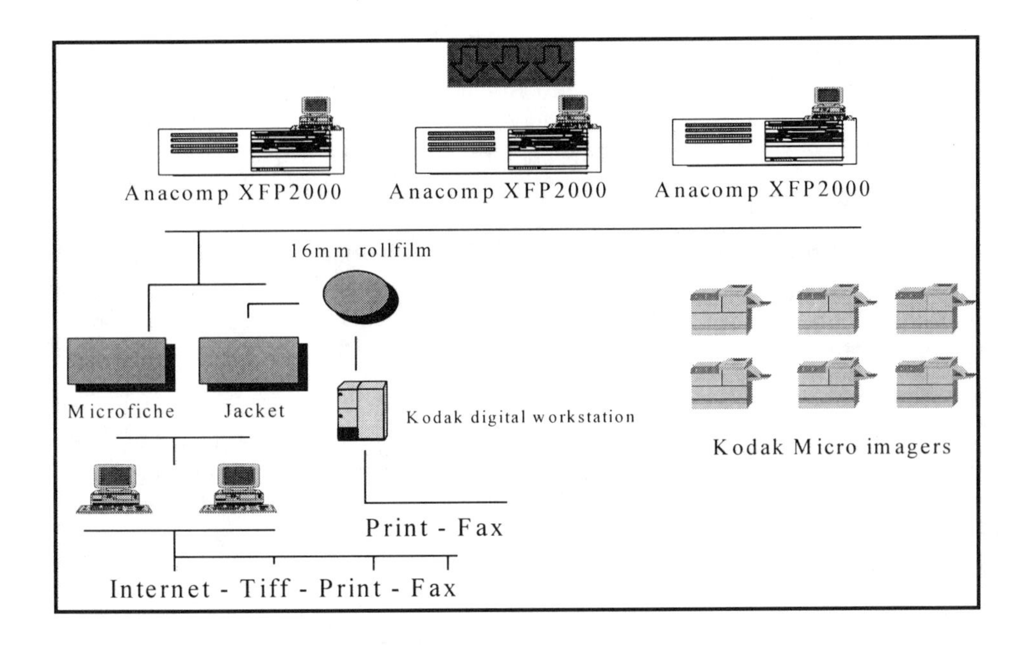

Document input/output management

In order to dispense with all paper information (files and computer output) as quickly as possible, all computer data are put directly onto 16mm rollfilms using Kodak cameras and onto image-fiches on Anacomp's XFP2000 COM system. Electronic documents can always be created from these micrographic information carriers.

Those divisions of the organization which have no necessary infrastructure have time to prepare their infrastructure. The fiches take up little space and form a usable back-up for files. The micrographic carriers can be scanned and entered in the image system as Tiff-files. Filling the electronic archive with the existing paper output is a manageable task and also a necessary one. Once a part of the organization has its infrastructure available, all that has to be done is to read in tapes containing the scanned documents as images with data, and with indexes, into the system. The division will have a complete electronic system, in addition to a back-up on one of the micrographic carriers. Until then, it is easy to supply information through the usable fiches.

Retrieval on demand and off-line storage

From a whole jacket even a single image can be accessed on demand and the Tiff image can be resent via the Internet for example. Duplicates of the fiches are available for the various departments as a shadow-archive whenever they do not have the required infrastructure in place. Once a document is no longer in active use, that document can be stored off-line and once the document needs to be used again, it can be automatically re-entered as an active image.

1.C Employers Insurance Administration Department-EIA

The Employers Insurance Administration (EIA) Department took the decision to reengineer the process of registration, premium calculation and file handling. This department serves more than 280,000 Dutch employers for the insurance of their seven million employees/beneficiaries. The amount of collected information over the recent decades is used to track each individual worker in the Netherlands and thus calculate his/her taxable social security premium. The procedures to control that information are a vital part of the business process.

Previously, this Department handled over eight million filed documents manually. They had major problems due to the amount of paperwork involved, information was not easily accessible, files were often lost or misplaced, and thus premium processing was delayed. Furthermore, the department had no adequate means to manage those complex processes.

Work Management process

Gak's staff and customer service representatives use Wang's OPEN/image software to scan approximately 900 documents per day including correspondence, income statements, etc. PC DOCS-DOCS OPEN, manages the process of indexing, cataloging, processing, managing and retrieving these incoming documents as well as Gak's eight million on-line files.

The staff can locate a document anywhere on the system at any time improving responsiveness and productivity. They can retrieve files and images via PC DOCS–DOCS OPEN based on certain criteria and view the images in the OPEN/image software system. Despite the high volumes involved, the system gives users immediate access to files.

Wang's OPEN/workflow software is used to route the process information, through the system electronically to different areas of the department for further handling and investigation. This application also assigns tasks to the Gak staff members, alerting them automatically of any outstanding actions that need to be taken.

Links between DCC and the EIA Departments

Controlling unstructured information objects (image) documents were captured as described in section 1.1A-1.1C. The unstructured information is then needed to be indexed to serve the Employers Insurance Administration Department's needs.

The paper bulk-to-image conversion, as provided by the DCC department, then could be taken over by Wang's OPEN/image software batch utility which generates indexes for PC DOCS–DOCS OPEN indexing process. This indexed and structured information forms the core base of the Employers Insurance Administration Department's electronic document archive, stored on jukeboxes (800Gb).

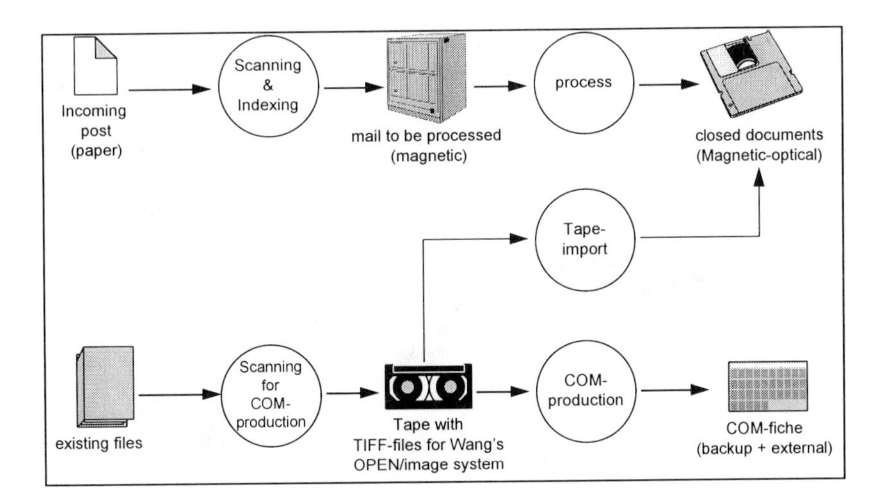

1.D Document Management

With the introduction of Wang's work management applications (imaging, document management and workflow), structured index information stored in a database facilitates fast access and relational on-line files.

Security of the files is guaranteed, because PC DOCS–DOCS OPEN controls the user and case accessibility. The document management system facilitates integration with other industry applications, thus controlling outgoing correspondence created by a case operator and consequently indexes the correspondence automatically prior to pre-defined HSM archiving.

An important aspect of this integrated solution is the use of structural components based on industry standards, in order to integrate various upcoming applications in use with the Wang OPEN/workflow for AIX, Wang OPEN/image for Sun and PC DOCS-DOCS OPEN for NT system.

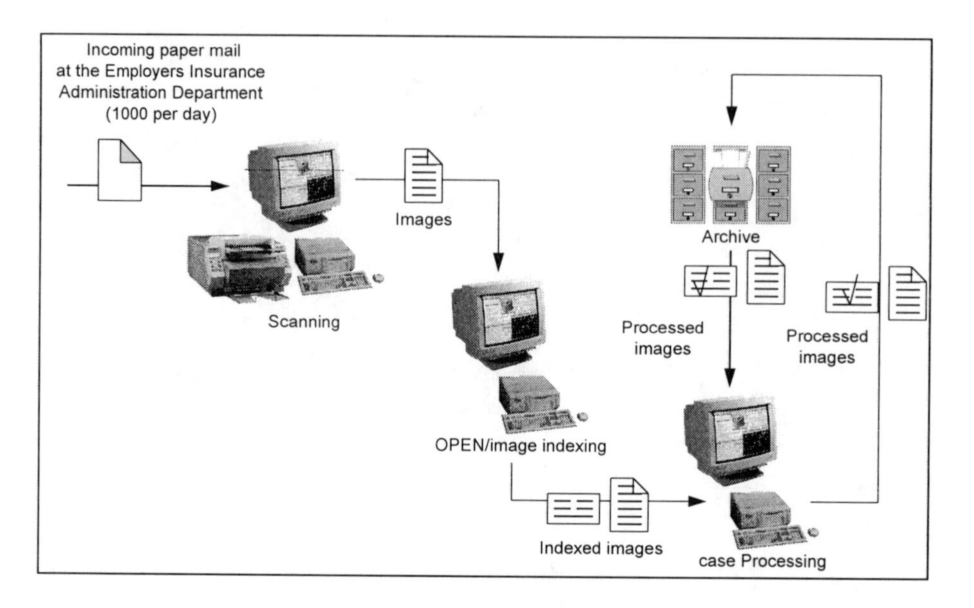

1.E. Process management/Reengineering

Two individual process management systems (data logistics and workflow) were introduced and implemented on top of the document and imaging systems. The OPEN/workflow application from Wang could easily be used as a steering instrument of incoming cases to be processed according to the case criteria inclusive of the steering of the related (image) documents stored. The data logistics concept, as developed by Gak, was integrated with the workflow component in order to introduce a modern and highly innovative administrative working method.

Key to the reengineering effort prior to implementation was Gak's decision to model based on industrial logistics processing principles. Gak selected BAAN software (also utilized by Boeing to manage their manufacturing-based industrial logistics), as the software to provide the logistical principals to support their administrative processes. In conjunction with the BAAN initiative, Gak also developed internally their own business modeling and process tools which they then integrated as a front end to the BAAN logistics software for their reengineering initiatives.

Two teams were then assigned; the Document Conversion Center (DCC) reengineering was determined by a 5-10 member team (depending on which aspects of the operation were being analyzed) of Anacomp, and DCC personnel with ASZ as the system integrator. The data logistics and work management/workflow initiative team consisted of personnel (upward of 25 at a given time dependent on processes being analyzed) from the Employers Insurance Administrative (EIA) Department, and Wang with ASZ as the overall conceptualizer and systems integrator.

As a result, users are freed from login into multiple systems and applications needed for case control. The system launches various applications, either business-critical or desktop applications and are architecturally controlled. In fact, it is no longer the user who determines the file position and its relational application, but the combination of Wang OPEN/image software, Wang OPEN/workflow, BAAN, and PC

DOCS-DOCS OPEN which perform those tasks automatically. The user/operator concentrates his/her activities via predefined work queues with priorities given by the workload balancing mechanism of OPEN/workflow software. Cases are automatically routed through the organization once case items are concluded. The whole business process therefore can be managed and maintained. A comprehensive set of management tools provides management information about case duration times, costs per task etc.

Section 2. What were the key motivations behind installing the system?

2.A Dutch Government Privatization of the Social Security Program

The Dutch Social Security program is both complex and changing dramatically. One of the many changes was the key motivator for Gak to implement the work management system. Previously employees were insured through their employer without any registration of the individual employee. For premium calculations the 200,000 employers informed Gak about the total amount of employees and wages and premiums were calculated as percentage of that total amount. Today every person in the system is managed individually and an insurance policy is issued per person.

This move represented a significant impact on Gak. The number of customers they needed to service grew from the collective program of approximately 280,000 employers, to one where, at the employee level, it increased the number of customers to be serviced to approximately 7,000,000 employees and beneficiaries of these employers for Gak.

This increase in customer numbers also exponentially increased the number of telephone calls and documents to be processed and managed by Gak. Calls for information average 1,400 daily with peaks up to 2,000 per day. The call desk is now able to answer most questions on the first call by utilizing the access to information available through the work management solution, eliminating the cost and expenses of a call-back. In addition, the level of correspondence increased to more than 1,000 per day requesting clarity on the changes to the system as well as modification and changes to individual accounts.

Through the automation of the work management technology and the reengineering of the processes, Gak now has a diverse work force that can respond to workload balancing based on peaks in demand, while downsizing the EIA department by approximately 50 percent from 240 to 125 today, with efficiencies gained by the system allowing for further streamlining of the employee pool. Gak has realized these staff reductions through the work management technology and process re-engineering with the data logistics concept even as their customer base grew from 280,000 employers to more than 7,000,000 individual customers.

Automation has also reduced the informational workload of Gak's customer base as all information and data are automated, reducing the informational requirements from several pages in the manual processes to approximately two pages today.

Business Process history:

The Gak organization, as any administrative organisation, records an extensive paper-based archiving method decades-long per individual. Since inception, Gak produces an enormous paper mountain. Information is collected, used and archived on paper. The Gak products to its clients (branch organizations), and to employers and beneficiaries are usually presented on paper. Since the introduction of IT methods, the use of physical files has increased even further. Millions of documents are collected and distributed every year and stored in files. The information within all these documents is hard to access for all departments needing to use that information.

Pre-project working methods

In the old situation, paper files had trivial limitations, only one user could keep a file in its possession for processing. The administrative tasks therefore were strict, none other than the case operator had access to a particular file. Interim reports to the customer as to the status of a case could not be provided, leading to tedious long processes and a lack of proactive customer service.

Problem definitions

- A significant amount of paper files was untraceable
- Files or parts thereof were lost
- Unsecured and unstructured
- No tracking possible as to contents completion
- Ranking order (dates of correspondence) within a dossier did not exist
- Misfiled contents (wrong case in wrong file)
- Frequent duplicates and duplicated version of file contents, leading to massive increase of paper use.
- Problems with dissimilar information resources; paper, film, tapes, microfiche

Project objectives

In order to overcome the above-mentioned bottlenecks, Gak defined their objectives to deal with the ever-increasing limitations of accessibility to documents and the information therein.

- Accessibility to various media types, magnetic and paper
- Research the possibility of introducing an image-based system
- Reduce costs by implementing an image-based document management system
- Structure work processes via a logistics and workflow management instrument.

Section 3. Please Describe the current system configuration: number of workstations number and type of software servers, scanners, printers, storage devices, number of locations involved.

Section 4. How is this system integrated with the company's other information processing systems?

Section 5. What stage of development is each part at; what has been installed?

These issues have all been covered above. All aspects of the implementation discussed in this Case Study are installed and in full deployment.

Next step in the planning stages is deploying the system to the 30 District offices throughout the country.

Section 6. Describe how the company has been impacted by this system. Be as specific as possible. A.) What costs savings or increased revenues have been realized since the system was first installed. B.) What productivity improvements have been realized. C.) How has the business problem been affected (compared to before the system implementation).

Currently the department processes some 280,000 files electronically per year, each approximately five to seven pages thick and growing at a rate of 25,000 per year. As a result of implementing redefined logistics in an administrative process and by integrating Wang's work management software, all relevant information is available immediately to the customer service staff, thus improving their responsiveness and productivity. Processing time has been reduced from one month to two weeks. Furthermore Gak Nederland saved two floors of office space, eliminating paper document storage. The twin goals of cost reduction and improved efficiency were met.

With the combination of all involved elements in this project the administrative process has brought Gak significant business benefit and cost savings.

- Reclaimed two floors, 1,600 square meters representing approximately 500,000 guilders per year, now occupied by office personnel
- EIA Department staff reductions of approximately 50 percent
- Total cost reduction of 25 percent representing 25,000,000 guilders per year
- Product quality
- Effective implemented process
- Efficient process flow
- Fast information supply
- Measurable cost structure

Epilog

With the introduction of data logistics concept and work management at Gak Nederland b.v., considerable changes occurred in work processes. In the pre-imaging era, incoming mail was distributed over various different task departments, each of them took care of further distribution and sorting of cases to users.

With the initial introduction of imaging, these time-consuming tasks were integrated in a reception office, where daily input with the use of high volume scanners, could be sorted and processed faster, thus leading to a seamless approach to work management.

Section 7. Describe the implementation process and methodology, the project team and any change management and business process reengineering issues addressed.

7.1 Project participants

- Gak Nederland b.v.
- ASZ automatisering sociale zekerheid b.v.
- Wang Nederland bv

- Anacomp B.V.

7.2 Project implementation electronic archiving, work management and logistics

We moved from dissimilar media types to an industry standard format. Due to the magnitude of the organizational paper-based archives, two separate projects were coordinated. These projects were started with two joint teams.

- Gak's Document Conversion Center and Anacomp and
- Gak's Employers Insurance Administration Department, ASZ, and Wang

The specific task results were that more than one medium was needed to meet the whole range of retrieval and storage needs. While Anacomp and DCC took the lead in converting the paper-based bulk and media conversion, Employers Insurance Administration Department, ASZ and Wang concentrated their efforts on the actual data logistics system and work management applications. ASZ acted as the overall systems integrator.

7.3 Data Logistics and Work Management

The Gak Group developed a data logistics concept defining four layers: strategic, tactical and operational management and the operational process itself. ASZ implemented a tactical steering layer with BAAN (Triton) industrial logistics software. The administrative process was reengineered as an industrial process: the products to be delivered, the raw materials needed out of stock or to be purchased and production. The result was a reduction of dataflow, simplification of processes, and effective production management. ASZ and Wang implemented Wang's OPEN/workflow and OPEN/image software and PC DOCS–DOCS OPEN document management as operational instruments. ASZ integrated those components with existing legacy systems.

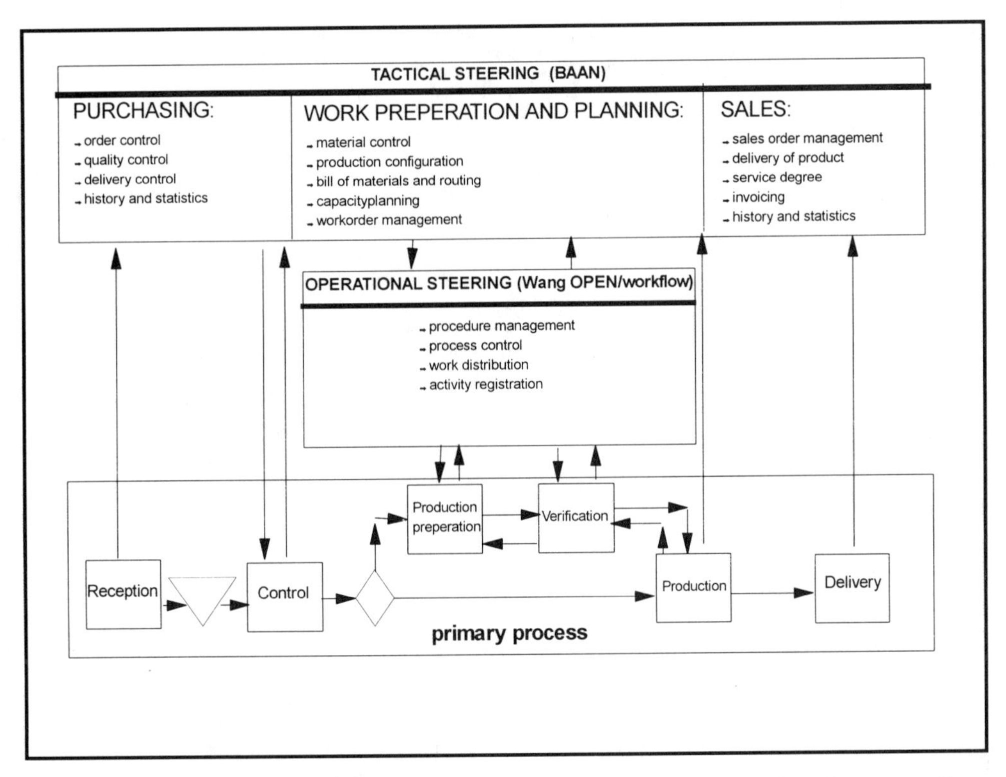

Section 8: The following highlights some of the detail included in the preceding Case Study on Gak Nederland b.v.,

Corporate Overview	• Gak Nederland b.v., • Largest Social Security provider in the Netherlands • 13,500 employees • 30 District Offices throughout the country • Serving 280,000 Dutch Employers, • Representing seven million employees and beneficiaries • Accounting for approximately 65 percent of Dutch workforce • Distribution of $11 billion per annum
DCC Project Team	• Document Conversion Center (DCC) Dept. • Anacomp and DCC Document Conversion Team • ASZ systems integrator

	• Multi-media conversion effort • Microfiche • Jackets and 16 mm film (CAR) • Rewriteable MO disks to store image documents • Foll-films for feeding digital workstation • CD-R for recording bulk mainframe data • Separate OMR and OCR projects • Scanning of microfilmed documents to TIFF files
Project Team	• Employers Insurance Administration (EIA) • EIA, ASZ and Wang • ASZ overall conceptualiser and systems integrator • Datalogistics and work management applications
Reengineering	• Data Logistics concept, defining four layer strategy as a tactical steering layer with BAAN (Triton) industrial logistics software • Strategic • Tactical • Operational management • Operational process itself
Work Management Platform	• 125 Users • Wang OPEN/workflow on IBM/AIX • Wang OPEN/image and Oracle 7 on Sun Solaris • PC DOCS OPEN Docs on Windows NT Server
Data Logistics Platform	• 15 management users • BAAN on DEC Alpha
Document Conversion Platform	• 16 mm rollfilms using Kodak cameras • Image-fiches onto Anacomp's XFP2000 COM System • Electronic documents created from the micrographic information carriers
Hierarchical Storage Management Platform	• Netstor HSM on Windows NT Server
Legacy Systems Platform	• Ingres on DEC VAX/vms

	• DBMS on DEC VAX/vms
Document Volumes Unique Correspondence	• 280,000 Files per Year • 6-7 pages thick • Growing at of 25,000 files per year
Document Volumes Financial Statements	• Seven million statements a year • 1 million on paper, 6 million on tape, disk, edi
Document Volumes Conversion	• 500.000 documents scanned per month • 250.000 historical microfilm documents scanned per month • eight million on-line document today
System Impact	• Total cost reduction of 25% representing 25,000,000 guilders per year • Reduced staff almost 50 percent from 240 to 125 in the EIA department • Eliminated 2 floors (1,600 square meters representing approximately 500,000 guilders per year) file storage, now occupied by office personnel • Processing time reduced 50 percent from one month to less than two weeks • Access to over 8,000,000 on-line documents for customer service representatives • Multi-media types used in the document conversion allow them to deliver the information to their external clients in the form in which they can use them. Internally provides access to those departments to yet implement the infrastructure for the work management system
Wang Software Selection Criteria	• Wang has a comprehensive architectural concept. • The open architectural concept has allowed Gak to select best of breed and price/performance by software application on the hardware platform, thus the Wang OPEN/workflow for AIX and OPEN/image for Sun, PC Docs/NT, BAAN/DEC Alpha, and Netstor/NT • Their core business is workflow and imaging which they know best • Software functionality best met the end user demand
Next Stage	Distribute system to the 30 District Offices throughout the country

Italian Chambers of Commerce Societa' Consortile di Informatica delle Camere di Commercio Italiane per Azioni (Infocamere) Padova, Italy

European Excellence Award: Imaging, Gold

Executive Summary

Societa' Consortile di Informatica delle Camere di Commercio Italiane per Azioni (Infocamere) is the Information Technology (IT) service company of the Italian Chambers of Commerce. Its role is to provide the 103 Italian Chambers of Commerce with IT solutions to help them manage their activities and their growing number of responsibilities.

Until 1994 this was, among others, the role of Cerved S.p.A. Since then Cerved decided to concentrate on selling business information collected from the Chambers of Commerce; Chambers created a new organization dedicated to providing IT services to them. This organization took the name of *Societa' Consortile di Informatica delle Camere di Commercio Italiane per Azioni* (short name, Infocamere). This company employs 700 people generating approximately $110 million each year.

Section 1. Describe the system application. What the system is used for, who are the users and what the job entails. How often or how many hours is the system in use on a daily basis.

1.A Company Background:

Wang Laboratories Inc., (Wang) began its relationship with Cerved/Infocamere began during 1993, when the Italian law reshaped the role of Chambers of Commerce and emphasized their responsibilities for collecting and storing official business documents from the companies, such as balance sheets and other records. This led Cerved to consider imaging technology and to evaluate a number of vendors, including Wang, FileNet and Plexus. They then selected Wang's OPEN/image™ software to implement their new 'BOND' system, Cerved's first large scale imaging project. Based on the good feedback they had from this first project, the customer decided to include OPEN/image software into the platform for their mission-critical, highly visible *Registro delle Imprese* project, whose implementation has been Infocamere main activity for the last two years.

In all its operations with Infocamere, Wang acts as a supplier of the customer's prime contractor and platform provider, Hewlett-Packard. A description of 'BOND' and *Registro delle Imprese* projects follows.

1.b 'BOND' project.

The application. BOND project was the first Cerved/Infocamere project based on a Wang's software platform. Along with OPEN/image software, Wang provided design and development of the UNIX-based imaging application. Designed, developed and installed during the first half of 1994, the system went into production on October 1994. The system is based in Padova, near Venice, at the Infocamere main computing center.

The customer's goal was to store as image documents the balance sheets of nearly 550,000 Italian companies. According to the Italian law each company must file every year its balance sheet with its local Chamber of Commerce, which is responsible for archiving all company records and making them available for public access.

With BOND, Chambers send to Infocamere a digital copy of each balance sheet they receive. Through a country-wide wide area network, users can access remotely the system, working both from the their own premises or from a Chamber of Commerce. Requested documents can be delivered through a number of distribution channels, including fax, mail and file transfer. This dramatically reduces the time needed to get these documents, avoiding the need for long, paper-based procedures at the Chamber of Commerce of the city were the company is based.

Therefore the system addresses the needs of everyone who is interested in the activities and in the financial status of an Italian company.

System technical description. The systems is based on a HP 9000 800 server. It is aimed to load and then distribute about 550,000 documents per year, each representing the balance sheet of an Italian joint stock or limited liability company. These balance sheets are compiled according to the new European Union regulations and provide in-depth information on the company activities and financial status. The average number of pages for each balance sheet is 22, but in some cases large corporations can submit balance sheets in excess of 1,000 pages. Each document is further subdivided in about 10 sections. Hence about 10,000,000 of pages are loaded on the system every year; Cerved/Infocamere plans to keep at least three years on line. Therefore by the end of this year the system will manage more than 30,000,000 pages.

Document capture is handled through five scanning centers, located throughout Italy, and equipped with an OPEN/image software application developed by the customer. These centers collect the balance sheets from the Chambers of Commerce, and scan them into 2GB UNIX DAT tapes. These tapes are then sent to Infocamere for loading; together with images they include, indexing data which allow the document to be cataloged in the system database.

The BOND system is currently able to load and index more than 7,000 pages/hour. This performance level is critical to the customer as most balance sheets are received during three months of the year and need to be processed without buffering, to have them on line immediately; therefore, input peaks as high as 100,000 pages/day must be handled in some cases. Loading and indexing is handled through a set of UNIX background processes which read the tapes, verify that their structure is correct, select random samples of documents which are then checked by operators against image quality and indexing data accuracy. If these checks are OK, image documents are moved on optical disks and indexed on Oracle, using OPEN/image Document Manager software services.

Since Infocamere corporate databases are currently kept on hosts, indexing data are also exchanged (as flat files) between the UNIX application and host to keep the two environments synchronized.

To post requests for balance sheets, users access remotely an IMS/DB2 host-based application which allows 'navigation' through Infocamere corporate databases. Data held on these databases which provide a rich set of information on all Italian companies, are used as 'indexes' to select the balance sheets required. As a result of their query users are allowed to define a 'request' which may include any combination of documents or document sections; also one of the available document distribution method must be selected.

IMS then passes these requests to the UNIX server where they are queued for processing. On the server, a set of background tasks reads the requests queue and schedules their execution, based on the request priority and on the resources available on the server. As a result of the execution, documents are faxed or printed; billing data are also calculated and sent to the host for invoicing. Currently the system handles more than 1,000 user requests per day, delivering as many as 30,000 pages/day. Some of these requests are delivered by fax; others are printed and then mailed. Recently new services have been added which allow for file transfer to Chambers of Commerce over the high speed "C-LAN" network (see 1C for details), and for delivery through email.

In addition, Cerved/Infocamere has developed its own application to download selected groups of documents to CD-ROMs, which are then shipped to the Chambers of Commerce to allow access to local documents.

A set of application programs running on both UNIX and host, based on SNA LU6.2 communication services, takes care of handling communications between the two environments. On the UNIX side, this application is based on HP's SNAPlus software.

All the applications, the database and the image server software are executed on the same UNIX server. The BOND system works non-stop, 24 hours/day, seven days/week (except for about 18 hours/week that are devoted to system maintenance). Please refer to section 3A for the description of the hardware and software platform on which the system has been built.

1.c Registro delle Imprese project.

The application. With law number 580 issued on January 1993 the Italian Parliament decided to broaden the role of Chambers of Commerce, making them responsible for collecting and storing all the information that companies must provide to government and to the business community. All this information will form the so-called *Registro delle Imprese* (Company Register), which will become the official repository of data and documents on all organizations which are involved in business activities in Italy (an estimated amount of some 4.5 million business entities).

Implementing this project is currently Infocamere's main responsibility. It has been estimated that the overall cost of this three-years project will be about $60 million.

The *Registro delle Imprese* is based on a network of systems, one for each Chamber of Commerce; each of these systems manages data and image documents about companies working in the Chamber of Commerce geographic area. A central system located at the Infocamere computing center in Padova will collect and merge data and images from all the Chambers, thus acting as a back-up site and providing access to countrywide information. Managing image documents is an important part of *Registro delle Imprese* applications. The estimated amount of paper to handle is about 30,000,000 pages/year, and the project's requirements are to keep these documents on-line for many years (the actual number of years depends on the kind of document).

To implement scanning, indexing, archival and retrieval functionality for image documents Infocamere has used Wang's Open/image software to develop a subsystem called EIDOS. Currently EIDOS is installed at 50 Chambers. The installation process is still going on and will be completed by the end of the year, as law n. 580 requires the project to be running in production at all Chambers by January 1997. *Registro delle Imprese* project milestones are summarized in the following table:

December 1993	Law 580 is approved by the Italian Parliament.
January 1994	Chambers of Commerce begin to act as unique point of collection and storage of business documentation submitted by companies. Paper documents are handled manually at this stage.
December 1994	Wang Laboratories Inc., signs a contract with HP Italy to provide OPEN/image software for *Registro delle Imprese* project.
December 1995	Pilot Chambers begin scanning and indexing documents. According to the law, documents filed since January 1994 will have to be scanned into the new systems.
February, 1996	*Registro delle Imprese* project, without image support, is running at all the Chambers of Commerce.
January 1997	Full *Registro delle Imprese* project (including imaging support) will be running in production at all the 103 Italian Chambers.

The technology. All the systems are based on HP 9000/800 servers, running HP-UX operating system, Oracle RDBMS and Wang OPEN/image software; client/server applications are developed using TI's IEF CASE tool and Visual Basic. OPEN/image software includes Server for HP-UX, Optical Disk Management System 5.25" standalone and jukebox, Print Server for HP-UX, API and Windows workstation for HP-UX, Fax for Windows and ALCOM LFR server, Cabinet for Windows, Custom Controls for Visual Basic. The customer has an option to buy other imaging software, including new products which could be released in the future.

The Chambers of Commerce and the Infocamere computing center in Padova are linked by the high-speed C-LAN private network. This network is provided and operated by Telecom Italia, Italy's TC services provider; it is based on frame relay technology and features access links ranging from 128Kbps to 2Mbps, according to each Chamber's needs. Padova computing center is connected through 6x2Mbps links. The network will support services such as remote system management from the Infocamere main operation center, and access from a single Chamber to the *Registro delle Imprese* central machine. In the near future other organizations will be connected to the network, to support transactions among the Chambers and other governmental offices as required by *Registro delle Imprese* procedures.

Infocamere is currently considering Wang Laboratories Inc., Laboratories' OPEN/workflow to manage in a consistent and controlled way the large quantity of cases that companies are filing with the Chambers to comply with the Company Register regulations. Please refer to section 3B for a description of the system hardware and software platform.

Standard compliance requirements. An important feature of *Registro delle Imprese* is that image documents stored on the systems have the same legal value of the original paper documents; Chambers will have the option to destroy paper and to rely only on digital imaging for all their activities. To guarantee that image documents cannot be altered and can be easily accessible for long time in the future, the Italian authority for governmental computing, A.I.P.A., has issued a set of rules to which imaging technology must comply. An important piece of these rules is adoption of ISO DIS13346 standard for the optical disk file system. Wang is working on this technology and plans to support it in future releases of OPEN/image software server.

Business value for Wang Laboratories Inc., Laboratories Inc. The following table summarizes the value of contracts Wang Laboratories Inc., Laboratories Inc., has signed with Infocamere for this project.

Wang Laboratories Inc. contracts value:
$1,500,000 (OPEN/image software)
$170,000 (Consultancy and software maintenance - 1st year)
$250,000 (Consultancy and software maintenance - 2nd year)

Future developments. In the near future, Infocamere is planning to implement the central machine of the *Registro delle Imprese* network, managing the countrywide repository for data and images collected from all the Chambers.

This countrywide image document repository will be fed by EIDOS systems at the Chambers which will use the C-LAN network to load remotely imaged documents into the central server. To integrate the central system with the Chamber's systems from a user's perspective, Infocamere will develop a query software which will be able to look for the required information on the local or on the central server, transparently to the user. This will allow the central system to act as a backup site for all Chambers and also to provide cross-Chamber image document access.

Since balance sheets are a subset of documents managed under the *Registro delle Imprese*, the new system will also merge BOND into the *Registro delle Imprese*. This will allow Infocamere to provide external customers with an unique access point to all *Registro delle Imprese* image documents, extending the service BOND is providing today to a much wider document database.

Section 2. What were the key motivations behind installing the system?

The institution of the *Registro delle Imprese* (something that was first drawn up in 1943 but was eventually defined and implemented only 50 years later) broadened and strengthened dramatically the Chambers of Commerce's role, making them responsible for collecting, storing and making available upon request all company records. The Register was also conceived as a countrywide repository, able to provide access to every document from every Chamber.

This made imaging technology a fundamental part of the project's technical platform. Without the services provided by an enterprise-wide production imaging system, it would have been impossible to make available across all the country the huge quantity of documents that this system will manage.

Section 3. Please Describe the current system configuration: number of workstations number and type of software servers, scanners, printers, storage devices, number of locations involved.

3.A BOND PROJECT System Configuration (Background & Application Description in Section 1.B), and System Description:

General system description:
Number of servers: 1

Number of workstations: about 20 directly linked to the system and used for faxing and system maintenance purposes. Customers can inquire the system from about 3500 workstations connected to Infocamere hosts.
Number of scanners: no scanners directly connected to the system. Scanning is handled through 5 systems that manage approximately two Bell&Howell scanners each; images are then transferred to BOND system via tape.
Number of printers: 2

Hardware platform:
HP 9000/800 I70 server
4x12" ATG jukeboxes, each with 4 10.2GB optical drives
5 HP Netserver PC as fax servers, equipped with INTEL Satisfaxtion & Gammafax Gammalink fax boards (total of 16 fax lines)
2 HP Laserjet IV Mx Si printers

Software platform:
HP-UX 9.04
Oracle RDBMS Server 7
HP SNAPlus 3.0
Wang OPEN/image software API for HP-UX 3.0
Wang OPEN/image software Windows workstation for UNIX 3.0, Wang OPEN/image software Fax for Windows 1.01
Wang OPEN/image software server for HP-UX 1.1.5 with 12" WORM Jukebox management software
Wang OPEN/image software print server for HP-UX 1.5
ALCOM LAN Fax Redirector 2.10 & 2.15

3.B Registro delle Imprese PROJECT System Configuration (Background & Application Description in Section 1.C), and System Description:

General system description:
Number of servers: 104 (104 different locations, one for each Chamber of Commerce plus one central server).
Number of imaging workstations: Overall the *Registro delle Imprese* systems include some 1,000 workstations. About 300 of them are initially enabled to access image documents (number of imaging workstations at each site ranges between 10 and 2).
Number of scanners: about 150 overall (number of scanners at each site ranges between 4 and 1).
Number of printers: 1 for every system.

Hardware platform:
HP 9000 800 "K" and "D" series servers
5.25" HP "FX" (with 2.6GB drives) jukeboxes (each server with one jukebox initially)
Fujitsu scanners with KOFAX scanner control cards.

Software platform:
HP-UX 10.01
Oracle RDBMS Server 7
Wang OPEN/image software server for HP-UX 1.3, with 5.25" WORM Jukebox management software
Wang OPEN/image software print server 2.0 for HP-UX
Wang OPEN/image software Custom Controls for Visual Basic 2.0, Wang OPEN/image software Runtime for Windows 3.7.4 (client environment is currently Windows 3.11, will be migrated to Windows95 over time). Kofax KIPP software used to manage scanners.

Section 4. How is this system integrated with the company's other information processing systems?

Infocamere information processing system is quite complex and includes a central host-based platform (IBM/DB2 environment) and a distributed UNIX-based platform (HP/Oracle environment); a central UNISYS environment is also running but will be migrated by next year. The C-LAN network physically interconnects these three platforms; databases on the three platforms can cross-exchange data.

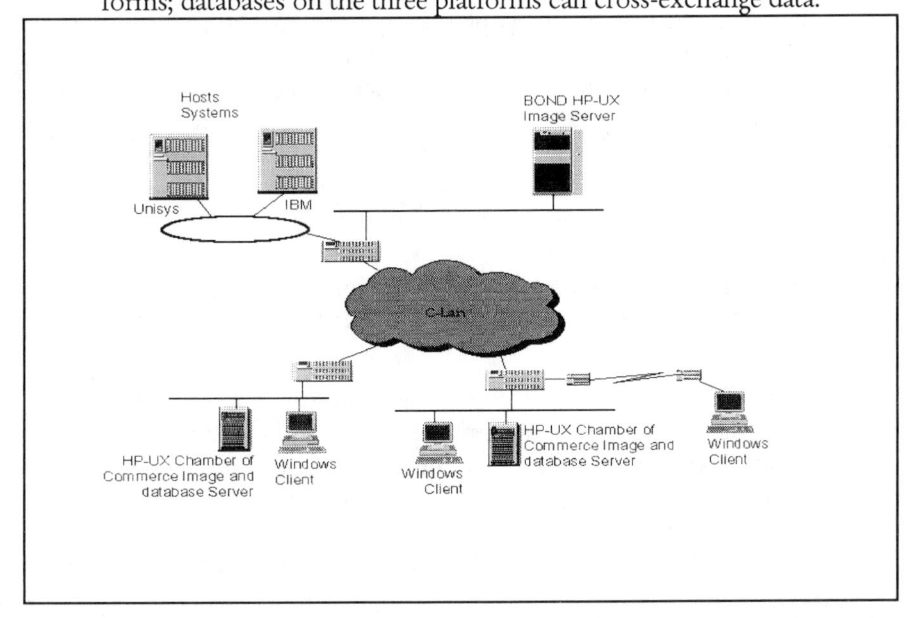

The imaging system is tightly integrated with this information processing system. All the image documents managed by both BOND and *Registro delle Imprese* are indexed through Oracle and then connected to the relevant data on corporate databases. This allows users of these databases to easily find and access image documents. An example of this integration has been described in the previous sections, showing how external customers can request balance sheets to BOND through the IBM environment which acts as a front-end for the imaging system.

The following diagram shows the current situation of information processing system at Infocamere.

Section 5. What stage of development is each part at; what has been installed? What is up and running? What is still in the planning stages?

BOND central system has been in production since 1994. The systems at the Chambers running EIDOS imaging application are being installed currently (about 50 in production as of September 1996). All the 103 systems at each Chamber of Commerce will be in production by January 1997. The central EIDOS system holding the country-wide image document database is still in a planning stage.

Section 6. Describe how the company has been impacted by this system. Be as specific as possible. A.) What costs savings or increased revenues have been realized since the system was first installed? B.) What productivity improvements have been realized? C.) How has the business problem been affected (compared to before the system implementation)?

The BOND system allows Cerved/Infocamere to offer to the market a service that previously did not exist. This system received some 1,000,000-user queries during 1995.

As to *Registro delle Imprese* system, it is a completely new infrastructure that is meant to offer new services to the Italian business community. The key advantage which the system is expected to bring is an unprecedented level of 'transparency' to the Italian business community. The system will allow getting in a matter of minutes all the information about a company's business, its management and its owners. Performing the same researches without the system and the associated imaging technology would have required days or weeks in the past. Law n. 580 has brought a countrywide standardization with full implementation across the Italian Chambers of Commerce.

Information on Italian companies may now be obtained at an individual's local Chamber of Commerce as opposed to requesting from the Company's local Chamber saving time and travel.

Section 7. Describe the implementation process and methodology, the project team and any change management and business process re-engineering issues addressed.

7.1 Project participants. The project has been entirely developed by Infocamere through their internal resources. Infocamere subdivided the whole project in a number of smaller projects, each managed by a dedicated team. A high-level management committee coordinates these teams. HP and Wang provided consultancy to project teams on system and application design.

7.2 Business process reengineering issues. The systems will have a very high impact on the way Chambers of Commerce perform their activities. However each Chamber is an independent organization, with its own specific organization and processes; how and to what extent the current situation will be impacted by technology is at the end a Chamber's decision. Infocamere will assist each Chamber in introducing this new technology into their organization, by providing consultancy and proper tools such as workflow technology.

Organizational Overview

Italian Chambers of Commerce	103 independent organizations, each covering a part of the country and providing business support services. Italian business entities supported by the Chambers of Commerce are approximately 4,5 million Impacted in 1993 when Italian law number 580 reshaped the role of the Chambers of Commerce Law mandated each of the Chambers of Commerce as responsible for collecting and storing ALL documents and information that companies provide to government and the business community. The institution of the *Registro delle Imprese* - 'Company Register' is a result of law n 580. Law 508 was a major change in defining Chamber's mission and role, and in promoting use of Information Technology in their activities. This led Chambers to establish in 1994 a new company - Infocamere - dedicated to providing IT solutions to assist them with their mission.
Infocamere	*Societa' Consortile di Informatica delle Camere di Commercio Italiane per Azioni (*Infocamere) is the Information Technology service company of the Italian Chambers of Commerce Infocamere's role is to provide the 103 Italian Chambers of Commerce with IT solutions 700 employees, generating $US110M/year Increased document requirements in Chamber's activity lead to imaging solution. First large-scale imaging project in 1994, called BOND (Balance Sheets); Wang OPEN/image software selected between FileNet and Plexus. Since 1994 Infocamere has assisted Chambers of Commerce with the implementation of the IT infrastructure for the Company Register program. A key component of IT platform for the Company Register program is production, enterprise-wide imaging services provided by OPEN/image software through an application subsystem called EIDOS.
Cerved S.p.A.	Until 1994 IT service provider to the Chambers was one role of Cerved Since 1994 Cerved focused on selling business information collected from the Chambers of Commerce. Chambers created Infocamere to replace services provided by Cerved.

Wang & Hewlett-Packard	Wang Laboratories Inc., Italy acts as supplier of the customer's prime contractor & platform provider, Hewlett Packard All imaging systems installed are on HP 9000/800 servers
Installation	Wang Laboratories Inc., provided design and development of the UNIX-based BOND imaging application Infocamere provided systems integration, application design and development, system management for the whole Company Register project (including EIDOS imaging subsystem); HP and Wand provided consultancy to Infocamere. BOND System placed in production 10/94. Company Register being installed in production during the second half of 1996; full production expected by January 1997.
BOND System Project @ Infocamere in Padova, Italy	First Cerved/Infocamere large scale imaging project, based on a Wang software platform— OPEN/image Software Cerved/Infocamere offered the market a service that previously did not exist— remote access to an image document database of balance sheets from Italian companies. Wang provided design and development of the UNIX image application Designed & installed first half of 1994, placed in production 10/94 System installed at the Infocamere main computing center in Padova System stores each year as image documents 550,000 balance sheets of Italian Companies. BOND receives the digitized copy of each balance sheet from the Chambers of Commerce, through one of five scanning centers. Host legacy integration provided to IMS/DB2 and Unisys host BOND system balance sheet documents are a subset of the documents to be managed by *Registro delle Imprese* ('Company Register') Project. The two projects will be then merged in the future.

BOND System Document Management & Volumes	Chambers of Commerce manage the balance sheets of nearly 550,000 Italian Companies. By law each company must file annually with its local Chamber of Commerce, who are responsible for archiving and providing public Access Average number of pages is 22 per company, some in excess of 1,000 pages with each being subdivided by 10 sections on the average. Approximately 10,000,000 pages are loaded on the system every year Infocamere plans for maintaining 3 years on-line Approximate 30,000,000 pages on-line end of this year 1996 A country-wide WAN (wide area network) allows users remote access via fax, mail and file transfer System has eliminated the need to travel to the local Chamber of Commerce where a company is based to access their financial files
BOND System Remote Scanning Centers & Document Loading	5 scanning centers located throughout Italy Scanning centers collect the balance sheets from the Chambers of Commerce Documents are scanned and indexed to the 10 subdivisions into 2GB UNIX DAT tapes Tapes are then forwarded to Infocamere for loading BOND System is able to load and index more than 7,000 pages/hour During 3 month peak of receiving balance sheets the BOND system manages up to 100,000 pages/day Infocamere corporate databases are host based (IMS/DB2 & Unisys) and indexing data is exchanged (as flat files) between the UNIX application and host to keep two environments synchronized.
BOND System Q/A Process	Loading and indexing are handled through a set of UNIX background processes which read the tapes, verify their structure is correct select random samples of documents which are then checked by operators for image quality and indexing accuracy Once passed for Q/A the image documents are moved to optical disks and indexed on Oracle, using OPEN/image Document Manager Services Software

BOND System Access	Customers make inquiries from approximately 3,500 workstations connected to Infocamere hosts More than 1,000 user requests/day, delivering as many as 30,000 pages/day Approximately 1,000,000 user queries ran against the system during 1995. Users access an IMS/DB2 host-based application which allows "navigation" through Infocamere corporate databases 'Request' may include any combination of documents and must identify preference for document distribution Requests are queued to the UNIX server for processing based on schedules and request priority Once executed documents are faxed or printed and billing data are calculated and passed to the host for invoicing Recent system service enhancements allow for file transfer through high-speed C-LAN Network to Chambers of Commerce, and for delivery through email Infocamere has also introduced CD ROM services, where they download selected groups of documents from BOND on to CD-R and ship to the Chambers of Commerce allowing for local access to documents
BOND System Availability	nearly 10-10 1/2 hours/day, 7 days/week Scheduled downtime for system maintenance

Registro delle Imprese ('Company Register') Regulation

Registro delle Imprese & EIDOS System Project	Drawn up originally in 1943 Defined and implemented 50 years later by law 580/1993 issued by the Italian Parliament. Identified a completely new infrastructure meant to offer new services to the Italian business community with an unprecedented level of transparency to this community Broadened and strengthened the role of the Chambers of Commerce, making them responsible for the collecting, storing, and access upon request of all company records Register originally conceived as a country-wide distributed repository, based on Information Technology tools, able to provide access to every data and every document from every Chamber Estimated cost of 3 year project is $US60 million Imaging technology became a fundamental requirement to provide this level of services and manage the huge quantities of documents enterprise-wide to the Chambers of Commerce Infocamere has implemented the EIDOS imaging system in each of the 103 Italian Chambers of Commerce to support the *Registro delle Imprese* Project through imaging.
Optical Legality/Italian Law	*Registro delle Imprese* identifies image documents stored on system with the same legal value of the original paper document Chambers have the option to destroy the paper Italian authority for the Italian governmental computing—A.I.P.A.— has issued a set of rules on imaging compliance which includes the adoption of ISO DIS13346 standard for the optical disk file subsystem, to ensure documents cannot be altered and remain easily accessible for well into the future. Wang is working on ISO 13346 and plans to implement it in future releases of OPEN/image server software.

EIDOS System Project & Geographic Distribution	A subsystem for scanning, indexing, archival and retrieval functionality - EIDOS - is installed in all 103 Chambers of Commerce EIDOS based on OPEN/image software running on HP 9000/800 servers; Oracle used as DBMS. Systems managed centrally by Infocamere. Law n. 580 requires ALL documents filed since January 1994 to be scanned into new system(s). Estimated amount of paper of 30,000,000 pages/year with multi-years' worth of retention on-line to system (dependent on type of document & legality). 50 sites are now installed and running Remainder to be installed by end of year Law 580 requires the full project, including imaging, to be running in all 103 Chambers by January 1997 Link provided by high-speed C-LAN private network provided and operated by Telcom Italia TC services is based on frame relay technology and features access links from 128kps to 2Mbps Infocamere computing center connected by 6x2Mbps links All EIDOS systems will be integrated with a centralized system at Infocamere centralized computing site in Padova, Italy that will act as back up to the data as well. EIDOS Central System Repository will be fed by the EIDOS systems from each of the Chambers utilizing 'C-LAN' network to load image documents remotely into the central server The centralized EIDOS system will eventually be integrated with the BOND System (Balance Sheets & Financial Information).

Future Direction	Near term Infocamere will implement the central machine of the *Registro delle Imprese* network This system will manage the countrywide repository for data and images collected from all Chambers Repository will be fed by the EIDOS systems at each of the 103 Chambers utilizing 'C-LAN' network to load image documents remotely into the central server. Query front end to be developed to locate documents on the central server or at the local Chamber transparently to the end user As balance sheets are only a subset of documents managed under the *Registro delle Imprese*, Infocamere will be able to provide a unique access point to all *Registro delle Imprese* image documents, extending the service BOND provides today to a much wider document base Additional organizations to be connected to the network, to support transactions among the Chambers and other governmental offices as required by *Registro delle Impress* OPEN/workflow software being considered to manage and control large quantity of cases

Martinair Holland
Schiphol Airport, The Netherlands

European Excellence Award: Imaging, Silver

Executive Summary

At Martinair, the second largest airline in the Netherlands, with over 2000 employees, the Revenue and Cost Accounting department processes coupons for accounting purposes and for billing tour operators and other airlines. The average number of coupons processed per year runs to about one million. The Martinair Input System (MAIS) is used within this department to replace the cumbersome and expensive process of manual data entry and the laborious archive process. With the MAIS system Martinair has realized significant process improvement and cost reductions, resulting in a return on investment (ROI) of less then 18 months.

The MAIS system combines Olivetti software for scanning, OCR, repair, data-entry and process management with an ImageTrac high speed color scanner (IBML) and FileNet storage and retrieval software. The MAIS system has been implemented by Olivetti in conjunction with the Martinair end-users and ICE project management. The pragmatic user-centric approach resulted in a production ready system within four months.

For other Martinair departments and processes, the MAIS system provides a solid and easy method to expand IT architecture for present and future developments in the area of imaging and workflow management. The extensive interest of other airlines and related companies is proving that Martinair, with the implementation of the MAIS system, is ahead of a lot of the competition.

The Challenge

Prior to implementing MAIS, a pool of data entry typists would key in the relevant data from the coupons. An arduous and monotonous task, leading to errors and resulting in highly demotivated employees. The department faced a 2-3 month backlog and a laborious archive process. The usage of temporarily staff was costly and inefficient. Whenever a customer called with a query, it sometimes took up to twenty minutes to retrieve the coupon.

Today, the MAIS system has reduced coupon backlog during peak periods to two days and provides instant access to stored coupons for answering customer queries while increasing archive retrievals to nearly 100 percent. MAIS resulted in highly improved quality and dramatic cost reduction. The total return on investment (ROI) was less then 18 months for the Revenue Accounting Department alone. For other Martinair departments the MAIS system has paved the way to use the FileNet imaging (& workflow) technology for further process improvement and cost reduction.

Olivetti, the system integrator, realized the MAIS system at Martinair is in direct cooperation with the end-user department. This user-centric approach was one of the key elements during the implementation and is reflected by the fact that MAIS is operational in the middle of the Revenue Accounting Department. The Olivetti software for scanning, OCR, repair and process management (together called Document Input System—DIPS) works in conjunction with an ImageTrac scanner from IBML and FileNet software for Storage and Retrieval. The usage of color images endorses the highly innovative concept of MAIS.

MAIS- Functional overview

The MAISsystem scans airline documents (IATA tickets), extracting airline codes, form and serial numbers and check digit. The scanned documents (JPEG format) are stored on optical disks with index fields based on the extracted previous data of the airline documents. The index fields at the moment are:

- flight date
- flight number
- airline code
- form and serial number
- document-id
- aircraft registration (1/1/97)
- from—to (1/1/97)

From a functional perspective, the MAISsystem replaces the old manual process with seven distinctive working steps. The usage of imaging technology almost implicitly also improved the workflow within the Revenue Accounting department. Working with the MAIS system consists of:

- Defining the Airline Flight to be processed by the system by flight number, flight date, aircraft registration, from-to and coupon count (which is done at the gate). With special separators the flight is split in economy class, star class, interline and post flight coupons. By defining the flight via the management station, this flight is now allocated and the system keeps track of this flight throughout the further process within MAIS
- Then the flight is scanned as a batch and counted. This count is compared with the coupon-count which was entered at the define flight (batch) stage. If there is a discrepancy the flight is re-scanned and the coupon-count is re-evaluated and if needed, adjusted. This coupon count is one of the various checks performed by the system to ensure that only correct coupon data and coupon images are used as input for the system.
- After scanning all coupons run through an off-line OCR process where the airline code, the form and serial number and the check digit are extracted. Due to color sensing and form-fix technology, only a very small percentage of coupons need to be completed on a special repair workstation.
- From the Oracle index database a corresponding data set is built for exporting to the IMPALA system (the KLM owned revenue accounting system that also operates as a host for Martinair. Prior to exporting, some

special type documents (± 10 percent) need additional data. These documents are automatically presented as an image on a PC workstation for this additional data entry.

- After export to and acceptance from IMPALA, all indexes and images are committed from the file server that was in control during the whole process, to the FileNet system (server and jukebox) for permanent storage on optical disks. The FileNet software allows the users at the Revenue Accounting Department easy access to and the electronic retrieval of the flight coupons and data for the remaining activities.

The MAISsystem is primarily based on the possibilities offered by the high speed ImageTrac scanner of IBML with its open, easy accessible, flatbed conveyor for document transport allowing easy correction in case of paper transport problems (staples, torn sides etc.). The ImageTrac scanner is fully software parameter driven, allowing Martinair to process different types and formats of documents in the same batch. By producing JPEG color output the scanner not only contributed to the quality of the process but also to the end-user acceptance. With this high speed color scanning capacity (the scanner can run up to 3 airline coupons per second) the backlog has been eliminated during the first months of production. During the summer of 1996 the system handled the full peak load in 4-5 hours a day.

The first phase users, in production since January 1996, are all working in the Martinair Passage Revenue Accounting Department. The total investment for MAIS therefore needed to be fully cost-justified by this department alone. After using MAIS for more then a year now, Martinair experiences benefits that exceed the original calculation.

The benefits of electronic storage and retrieval for related departments within Martinair are already proven and can be realized with a minimal investment. Therefore, in the following phase (1997) other departments will also get access to the MAIS system. Since the system does not need to be operational all day for the processing of coupons, Martinair is now in the process of investigating the usage of the MAIS system and components for other documents like Invoices, Airwaybills and personnel files. A third future development will be the scanning of flight-related documents and the logical foldering of those documents with certain coupons.

The key motivations for MAIS

Speed and efficiency improvement

Before the introduction of the MAIS system all data of the flight coupon had to be keyed into the IMPALA system. This was not only human and time consuming, it also carries the risk of all kind of data-entry errors. The OCR technology combined with several validations (like check digit) improved processing speed and quality.

Implementing MAIS reduced the backlog of processing flight coupons in flight envelopes of the summer season peak from 2-3 months down to 2 days. Following the reduction of the backlog, previous invoicing of tour-operators and other airlines (if applicable) was handled. An average gain of about six weeks is realized.

Reduction of human labor

The introduction of the MAIS system has resulted in a 50 percent reduction in staff. There is no longer a need for temporary labor during peaks which also reduces the need for training and extra supervision.

Reduction of archiving space

Before implementing MAIS all flight envelopes with flight coupons were archived physically. Now the archiving of 15,000 flight coupons (some 80 flight envelopes) together with the relevant index-data takes only one 5.25 inch optical disk. With the optical disk stored in a Optical Storage and Retrieval Unit (OSAR/Jukebox) model HP 120 T (88 slots) MAIS offers an on-line capacity of 1 year of flight coupons.

Retrieval of documents

With the FileNet component of the MAIS system, flight coupons can be retrieved within 10 seconds and the actual hit ratio of retrieving a coupon is nearly 100 percent. With the old physical "actual archive" close to the Revenue Accounting department, retrieving a flight coupon took approximately 15 to 20 minutes.

In the old situation every two months the flight envelopes in the physical archive had to move due to the lack of space in the "actual archive". After moving the archive several times it even became harder to retrieve documents.

MAIS—Configuration overview

The diagrams following provide an overview of the MAIS configuration. In the high level overview the three main components of the system are presented. Each component will be worked out in more detail separately.

High level overview MAIS

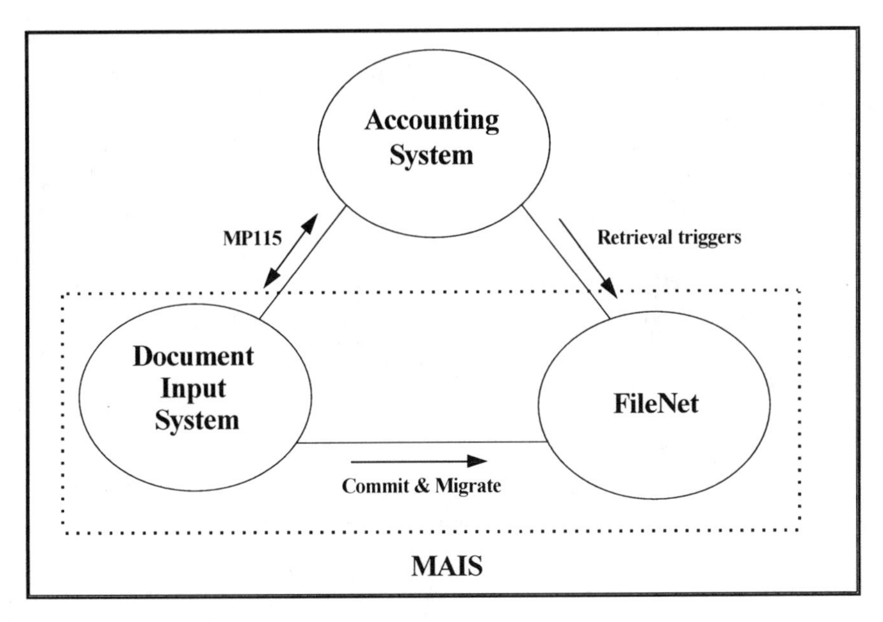

MAIS overview—the Accounting System:

The accounting system IMPALA is not really part of the MAIS system. The MAIS system creates a dataset that is exported to the IMPALA mainframe. The IMPALA application is a core application that is integrated in the Martinair financial system. The IMPALA system is used for validation. Correct processing of the exported dataset by IMPALA is acknowledged to the MAIS system and used for archiving purposes.

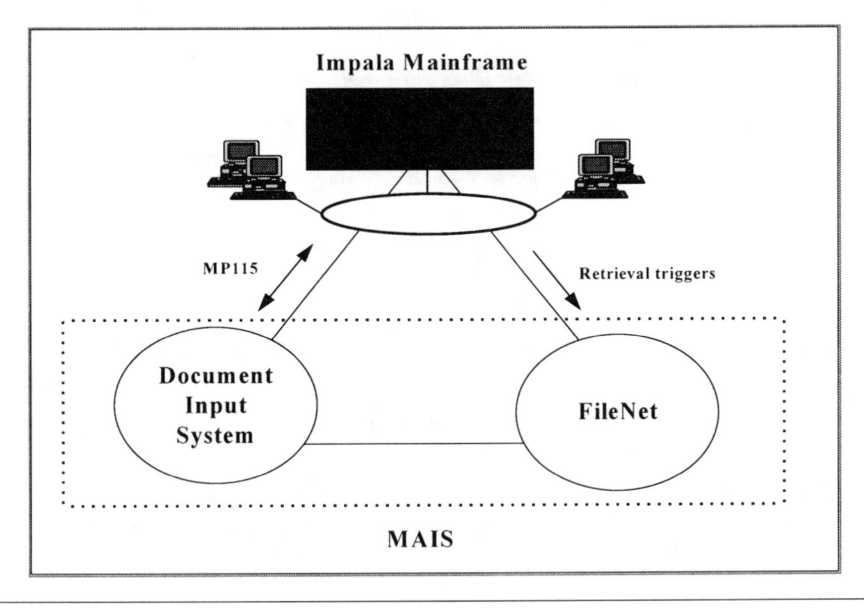

MAIS overview—the Document Input System:

The DIPS configuration contains several software modules concentrated around the high speed ImageTrac scanner. A more detailed overview is presented in the picture below. All workstations within the DIPS environment are connected via a separate Ethernet segment. This segment is separated from the Martinair network for performance reasons. The central database and file server (Oracle) controls the DIPS process. Via SQL statements this database is also triggered for management information output. After acknowledgment of IMPALA all Images and data are committed and migrated to the FileNet environment for actual storage. The DIPS environment supports the following processes:

- Define batch
- Batch (one flight) entry of coupons (scanning)
- OCR and Repair
- Batch management and control
- Additional data entry for special documents
- Committal and migration to FileNet
- Create export file

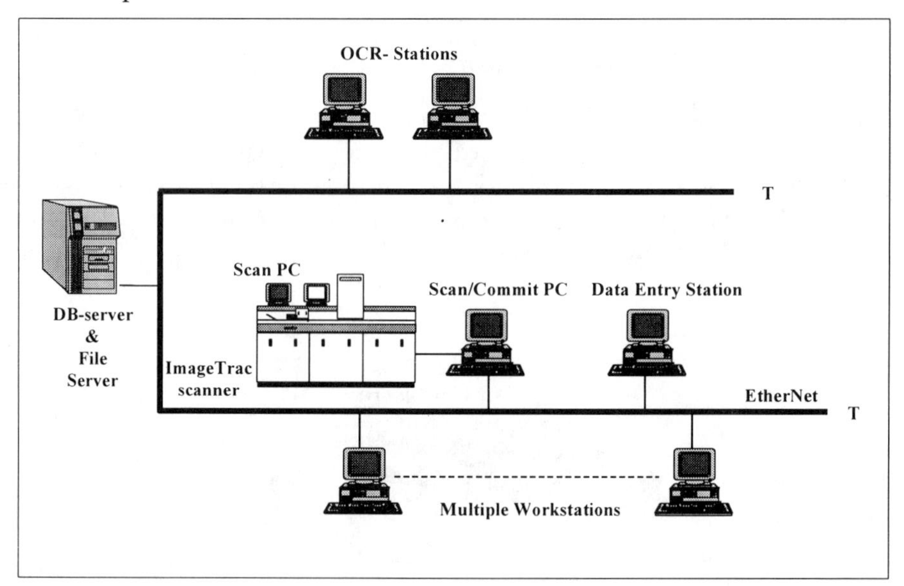

MAISoverview— the FileNet system:

The FileNet part of MAIS is used for the support of Storage and Retrieval of coupons and related data (indexes). Central in the FileNet system is the Image Management Server (IMS). Within the client/server-based concept of the FileNet software, the IMS supports access and retrievals on all workstations that are connected to the IMS. FileNet uses an Oracle database. Part of the IMS is the OSAR service that manages the actual storage of the colored images (and related indexes) to optical disk. The optical disks are stored in a HP model 120T OSAR (jukebox)

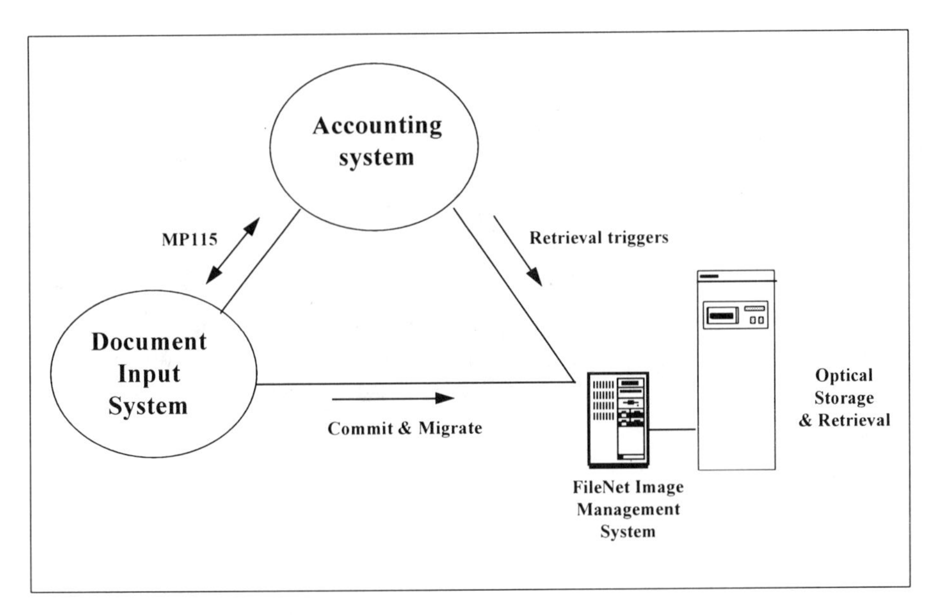

MAIS- Hardware & Software overview

The following hardware and software is located in the Revenue Accounting department of Martinair at Schiphol airport (one location). The whole configuration is installed and operates in full swing since January 1996 after a relative short development and implementation phase of 4 months.

The decision to build the system was taken after a feasibility study in July 1995 and after visiting both IBML in Alabama and an existing FileNet customer in the airline industry. The MAIS project itself started in September 1995 and was operational in January 1996.

MAIS Hardware

Scanner	IBML ImageTrac scanner (including basic unit, auto-feed unit, front JPEG image camera and single pocket stacker)
File— Database Server	Olivetti SNX 140, 64 Mb, 2x2.1 GB HDU, SCSI interface
IMS server	FileNet 6250 Base server, 32 Mb, 2x2.1 GB HDU, SCSI interface
Storage device	HP 120 T, 88 slots, 2x 5.25" disk drive
Scanstation	Olivetti M4-90 Pentium
Scan/Commit workstation	Olivetti M4-90 Pentium
OCR workstations	Olivetti M4-90 Pentium
Repair workstation	Olivetti M4-90 Pentium
Data entry workstation	Olivetti M4-90 Pentium
Retrieval Workstation	Olivetti M4-90 Pentium
Printer	HP Laserjet 4P

MAIS Software

Scanner	ImageTrac API
File—Database Server	Novell Netware / Oracle Workgroup
IMS server	AIX / FileNet IMS / Oracle
Storage device	FileNet SW driver
Workstations	• Oracle Workgroup clients • Olivetti software for Scan, OCR, Repair, Commit and Data-entry • FileNet Image display for Windows & FileNet WAL for windows

MAIS—Implementation

The project organization.

The MAIS project had a normal project management structure. On behalf of Martinair, Implementation Consultancy Europe (ICE) operated as project management. The Martinair (ICE) project manager worked with the Olivetti project team. The project team consisted of Martinair project management and the manager of the Revenue Accounting department, the Olivetti AE and technical consultant(s). The project itself had a very user-centric approach and low involvement of the EDP department. Due to this approach Olivetti was triggered for a turnkey realization and implementation of the system.

Development methodology

Based on a very pragmatic approach both Martinair and Olivetti agreed with the Rapid Application Development method (RAD) using prototypes for the different modules in the DIPS environment. The user-centric approach resulted in high involvement of the Revenue Accounting Department and easy acceptance of the end-users.

Due to the RAD method, the Martinair end-users could easy adapt to the changes that the imaging technology had on their daily work. The benefits became clear and the users were anxious to go life with the system.

Implementation

System installation, system testing and prototyping on the final production location, allowed high user-involvement and direct testing. At a certain date all documents were scanned, after testing thoroughly the whole system. The absence of errors in the parallel production phase reduced this period to less then a month.

The Martinair staff was trained by Olivetti consultants to work with the DIPS and FileNet components of the MAIS system.

MAIS—Future developments

- Within the current MAIS configuration a Rumba interfacing and 3270 emulation on the retrieval station is in progress. This interface will replace the current tape interface with Impala.
- More retrieval stations for the usage in other departments (like legal and claim processing) are part of a further implementation schedule during 1997.
- For the Revenue accounting department a feasibility study has been executed for the processing of non IATA tickets (charter tickets) based on forms and logo recognition. This addition to the MAIS functionality is expected to be operational during the summer of 1997.
- Outside the Revenue Accounting department other departments are interested to use the MAIS system for their documents (claim processing, invoices and cargo documents). A test with the scanning of invoices will start 1st quarter of 1997. The addition of cargo documents (airwaybills) is planned for the 2nd quarter of 1997.
- Another future development might be the usage of OCR/ICR technology to extract more data from the IATA tickets. This additional data can be used as input for management and marketing information.

MAIS—The impact on Martinair

As stated before, the ROI of the complete system including project costs will be within two years, based upon processing of IATA tickets (the current application). This ROI is due to the reduction of staff, less physical archive handling, space reduction and interest effects of early invoicing and related early collection of funds. If airway bills (cargo) are processed (2nd quarter 1997) even the ROI is estimated to come down to approximately 14 months.

The backlog is reduced to practically zero. Therefore the organization can be provided with accurate and actual load figures. This means a tremendous improvement in the quality of management information. Retrieval time has decreased enormously and the hit-rate has increased to nearly 100 percent. Besides these tangible improvements, there are several quality-related improvements.

In the former manual situation the whole process ended with the recording of the flight coupons. With the MAIS system all flight coupons are recorded at nearly the start of the process. As a result of this change of workflow there is a reduction of "document logistics" and "no-hit" situations.

Probably the best proof of the impact and benefits of the MAIS system on the Martinair organization is in the interest of other airlines and even companies outside the industry. During 1996 Martinair hosted several visits of other airlines.

Canada Institute for Scientific and Technical Information (CISTI) Document Delivery Division Ottawa, Canada

North America Excellence Award: Workflow, Finalist

The Challenge

The National Research Council's scientific library, Canada Institute for Scientific and Technical Information (CISTI) is the largest scientific library in North America with over 300,000 square feet of document storage. CISTI offers one of the world's best collections of journals, conference papers and reports in physical and life sciences, engineering, technology and medicine for its world-wide base of over 15,000 clients.

Traditionally, CISTI relied on photocopying paper copies of articles and conference papers to fill orders for specific documents. The paper copies were then delivered to the client using various mail and courier services, nationally and internationally. Approximately 20 staff members used 17 photocopiers to fill these orders.

In 1993/94, CISTI received over 330,000 orders for documents or over 1,100 orders per day. (The volume has since doubled.) From this over 244,000 photocopied documents were supplied resulting in delivery of almost 1,000 documents per day. With the average document consisting of 10 pages, CISTI delivered approximately 200,000 pages of requested materials monthly. CISTI projected that during the 1994/95 time frame, the daily number of documents delivered would have to increase to 1,200 documents supplied per day to keep up with the current information demand.

CISTI's Document Delivery operation was spread over six floors with the paper orders traveling from floor to floor during the course of the day. Once in process there was no way to isolate a specific order in the identification, document selection, photocopying and delivery processes.

CISTI realized that they would need to expand the range of ordering and delivery options and offers faster turnaround times on providing documents. To do this it was necessary to move into an electronic environment and offer electronic delivery options to clients. This necessitated moving from the traditional photocopying operation to an imaging environment and producing scanned images of requested documents that could then be sent electronically, according to the clients' preferred delivery method. The new electronic system baseline capability had to be able to scan 10,000 pages of documents and fulfill 1,000 separate document requests daily. The new system would also provide the basis for moving into an environment where documents could be stored electronically, copyright permitting.

CISTI also recognized that they would need to establish a new document order and delivery service business electronically via the Internet. With their extensive information holdings, CISTI was very concerned regarding the security and configuration of this groundbreaking new electronic commerce application.

Technology Challenges

- Set up the new imaging environment to scan images of requested documents that includes 13 scanning workstations distributed over five stack floors including the design and engineering of a new high speed imaging environment and system.
- Utilize business process reengineering to design an automated business process which would transform the current manual processes and accommodate a fully automated electronic document order, selection and delivery system.
- Design and develop a reliable and secure Internet order and delivery business process to establish a new electronic commerce line of business for CISTI. This entailed creation of an Internet based ordering system, where orders were received in a delimited file or standards based format for input into an SQL database, and an internet based transfer of order/client information from a mainframe to the IntelliDoc document management system.
- Ability to shift order receipt and document delivery from a paper based to an electronic environment, with resulting savings to CISTI and clients.
- Develop a series of MS Windows compliant interfaces to allow for documents to be automatically converted and sent out by Internet and fax, or to be printed at a high speed printer for courier delivery.
- Develop a system which would access information from several databases (catalogue of library holdings, client delivery preferences) in the course of order fulfillment.
- Develop a system to allow for the tracking of an order to any point in the workflow
- Combine many commercial "off the shelf" products to work with each other, and with CISTI's existing systems, to provide the most automated processing available, while providing the largest number of ordering and delivery options for end users.

Integration Overview

The principal business objective was twofold to provide a facility to allow the organization to automate manual paper based information gathering and dissemination and to establish a secure electronic information order, receipt and delivery business process which utilizes the Internet.

The process definition was based on a modified SDLC (System Development Life Cycle) oriented to process definition that was then used as the overall system definition. The FLOWMAN tool set was used to define and implement the electronic processes within the overall workflow process.

The integration included business process reengineering as well as designing new business processes required to support the system. The basic purpose was to provide the end user with worldwide desktop access to the CISTI information resources.

CANADA INSTITUTE FOR SCIENTIFIC AND TECHNICAL INFORMATION

The objective was to ensure that CISTI would not have to make changes to the systems already in place, while integrating the new enhancements, and allowing the integration of more packaged software as business needs changed.

As CISTI's holdings and the number of requests for documents continued to grow, and as end users asked for more options, it was essential that a new system be developed to allow for the expansion of CISTI's business. The new system would allow for the dissemination of documents to the widest possible audience.

The project team consisted of:

- Clare MacKeigan, Project Manager, CISTI
- Bill Ruane, Project Manager, System Architect, NSI Network Support, Inc.
- Michael K. Bennett, Project Manager, Logical Software Solutions Corporation

Product Selections

The key to success in this project was the selection of the right products for the tasks. Logical Software Solutions' FLOWMAN production workflow product and PSS' RIMS document management system were selected as the basic system in a Novell Server environment which utilizes SQLBase from Gupta as the database of choice.

The key attributes of both the FLOWMAN and RIMS products are as follows:

- A common SQL database that was not platform dependent, allowing FLOWMAN and RIMS to share data,
- External APIs for both FLOWMAN and RIMS allows easy integration with other MS Windows compliant applications;
- Open architecture that uses PC technology for the client functions and allows integration of the client desktop and database of choice;
- Utilizes rapid development tools for custom client requirements and system specifications

As described, the integration of FLOWMAN and RIMS was facilitated by the products' common open design philosophies. This architecture also allows integration of other products required to manage the document creation and delivery process.

These products included:

- WinFaxPro from Delrina
- Windsprint from Eureka Software
- SCO Enterprise server for Internet order receipt and delivery
- DIMS from PSS for scanner management and image creation
- SQLWindows for application interface development

All the software suppliers are leaders in their areas of expertise and have based their products on open system principles facilitating integration. This shortened the development for this complex system and allowed the team to concentrate resources on testing the new delivery methods and refining the supporting business processes.

Research and Development

The project team built a series of reproducible and configurable interfaces using messaging, OLE, and database access methods to link FLOWMAN for task sequencing to the communication interfaces needed for data capture and delivery to the various document management systems that are available. This approach provided information control, and the applications required for decision support services for automated document delivery by the CISTI staff.

The project team concentrated on the following activities:

- Interface and ergonomic design of user controls that included touch screen computer stations, high speed specially engineered foot pedal operated scanning station and scanner bed modification;
- Object-oriented design and programming for exchange of objects between common interfaces;
- Common indexing of workflow rules and information management holdings databases, and;
- Automated interface design between off-the-shelf communications packages (e.g. Delrina WinFaxPro) and the RIMS document management system.

Integration Results

1. Automated file transfer based on Internet with end-to-end messaging for application recovery of order/client information from a mainframe to the IntelliDoc server.
2. Automated interface between the document management database (RIMS) and the workflow software (Flowman) using common tables for workflow rules and order information.
3. Automated document delivery for a variety of interfaces including FTP, Internet, and Fax, including automated re-sizing, re-formatting and rotation as required.

Productivity Gains and Economic Achievements from Project

Productivity and Efficiency Dramatically Increased

First, the CISTI library has now gone from a manual order processing operation with only one method of delivery at a fixed time, to one which is fully automated and can process orders on a 24-hour turn-around basis.

For the first time, CISTI can now receive a constant stream of orders and process them on a continuous basis for a much lower cost per transaction. This allows them to receive normal orders to approximately 1:00 p.m. and still be able to deliver in the same daily cycle. Prior to this, an order would take up to *three or four days* to be processed. Also due to delivery time restrictions, the workforce only had a small four-hour window to process the daily order volume that hampered productivity.

The new electronic Document Delivery system is available 24 hours per day. It is connected to a host for 21 hours per day and delivers electronic orders based on a service type schedule. In order to facilitate the operation, a lights-out operation was developed for data base backups, order purging, etc. A significant effort has been made to make the client/server architecture both stable and "hands free" to reduce maintenance costs.

There are currently 8,000 member users served on a continuous basis both throughout Canada, and internationally. Primarily the users are located in North America, but there are some customers in Europe and Japan.

New Services Result in New Revenues

Second, the new electronic environment permits multiple levels of customer document information services for the library, which allows additional premium service options to be introduced into the regular processing stream.

Presently, CISTI is instituting a marketing effort targeted at expanding its international customer base, particularly in Europe and Japan. The new automated electronic systems and a delivery method developed by this system using the Internet makes this feasible now.

With management statistics now available from workflow, CISTI can now measure the effort associated with service and determine the most popular information sources requested by customers. This has already allowed CISTI to rethink the type and pricing of future services to be offered. This translates to increased revenues from premium service charges on certain information requests, increased productivity from the existing workforce, improved delivery services, as well as, a decreased cost for filling and delivery of normal orders.

With the installation of the integrated electronic Document Delivery operation, CISTI has embarked on a five-year plan that calls for a *doubling of annual revenues with no increase in personnel or costs* in a highly competitive international information supplier marketplace and in light of Canadian government funding cutbacks.

It is intended to extend the interface capability to allow the same process interfaces to the end user whether at work or in laptop/home environments.

Summary

Instituting Line of Business Workflow resulted in:

- Reduced delivery cycle time by 75 percent.
- Doubled production capability, by increasing production service to 24 hours per day.
- Instituted a groundbreaking new electronic commerce library and document delivery service via the Internet.
- Project an average annual increase of 20 percent in service revenues alone.
- Reduced international delivery costs by $1 per page ($10 per average length document), improved timeliness and improved quality of transmitted document by eliminating dependency on fax transmissions.
- Five-year business plan calls for 200 percent revenue increase overall.

Expansion of international client customer base.

New York City Office of the Comptroller
New York, New York

North America Excellence Award: Workflow, Gold

Executive Summary

New York City was experiencing record numbers of claims and contracts, and at the same time the resources allocated to processing them decreased. The New York City Comptroller's Office, with the assistance of its consultants Xerox and Universal Systems Inc. (USI), utilized imaging, workflow, and other client-server technology to enhance revenues and decrease operating costs.

NYC processes over 30,000 claims and incurs a quarter of a billion dollars in claim costs annually. Billions of dollars are spent on City contracts. The implementation of the Omnibus Automated Image Storage Information System (OAISIS) has significantly improved the work processes and productivity within the City's Office of the Comptroller to process contracts and claims. NYC plans on saving over $20 million dollars by the Year 2000 using Document Management Technology.

Innovative Use of Technology

The following are examples of how technology was used in OAISIS:

- The scan system used automatic page detection that alleviated scan operators from manually setting page sizes.
- Barcode technology was used to further automate the indexing process.
- Daily mainframe downloads were used to further automate contract indexing.
- A graphical outline control was deployed to provide an easy graphical method for indexers to manage large batches of images.
- The import fax server alleviated the need for scanning fax paper into the system.
- OAISIS integrated a custom payment system along with imaging and workflow that generated about $1 million in claim settlements per year.
- Softscan technology was integrated. Softscanning is the process of turning a document, such as a Microsoft Word document, into an image. This image is subsequently stored on optical disk and provides a permanent, unalterable record of the document. Integration of this technology removed the step of printing and then scanning.
- The OAISIS Wide Area Network (WAN) was created on the existing Information Technology infrastructure to support the bandwidth required for moving images across a network.
- The OAISIS file servers employed a hard disk system to cache optical images. This provided very quick image response times for the most frequently used images.

- Engineering size drawings were integrated with OAISIS. Documents larger than 11"x17" were imaged by the engineering scanning sub-system. Group IV Tiled TIFF was the compression format used.
- Photographs were integrated with OAISIS. Photos of accident scenes were stored along with the rest of the electronic claim folder. JPEG was the compression format used.

Degree of Complexity in the Underlying Business Process and IT Architecture

There were numerous organizations from NYC involved. They include:

- Bureau of Law and Adjustment
- Office of Contracts/Administration
- Management and Accounting Systems
- Financial Information Systems Agency
- Office of Management and Budget
- Bureau of Accounting
- Bureau of Labor Law
- Bureau of Information Systems

OAISIS was delivered to support all these NYC organizations. In addition, these organizations had sub-organizations. Because there were no business processing standards across them, the delivery of one seamless solution was complex. One of the many challenges that OAISIS faced was re-inventing and standardizing how NYC worked.

For example, within the Bureau of Law and Adjustment, there were seven divisions. Even though each division processed claims, each processed them differently. What Xerox/USI did with NYC management was create one master method to manage claims electronically. Therefore, the way a Personal Injury claim is processed is identical to a Property Damage claim.

OAISIS also had to deal with the existing IT infrastructure, such as:

- Enhancing the existing network to sustain the bandwidth necessary for images
- Educating NYC MIS personnel who mostly had mainframe backgrounds
- Educating NYC end users who were primarily familiar with mainframe or DOS applications
- Educating NYC MIS developers with client-server development technology

Advanced Workflow and Imaging Concepts

OAISIS employed Documetrix Workflow. Through the use of Documetrix Workflow the following automation was implemented:

- Automatic printing of Acknowledgment Letters. These letters are printed by OAISIS automatically as soon as a claim enters workflow.
- Simultaneous routing of documents to more than one user
- Automatic removal of disallowed claims which unclutter electronic workflow baskets
- Joining of additional documents in workflow and the subsequent notification to all users who has the claim/contract in workflow. Thus if a medical transcript is

sent by a claimant, the workflow system, in conjunction with OAISIS, will join them together and notify the personnel working on the claim.

OAISIS employed many imaging concepts. They include:

- Group IV TIFF - up to 8.5"x14" documents
- Group IV TIFF Tiled for large documents, such as e-size documents
- JPEG for photographs
- Network optimized for approximately 36,000 images/day
- WAN capable of supporting imaging
- Automatic page detection during the scanning phase
- Fax out and fax in technology; OAISIS converted from fax format Group III to Group IV
- OAISIS also supports the annotation of an image without physically modifying the original image. Annotations are stored as overlays to the image. Images can be viewed with or without the annotations. Annotations can be secured, prohibiting users from creating, modifying or viewing annotations. Printing of images with annotations is fully supported.

OAISIS supports the following types of annotations:

- *Redlining*, to create lines to point to areas on the image that require referencing
- *Highlighting*, to highlight an area of the image that requires attention
- *Post-it notes*, to create "sticky" notes that can be positioned anywhere on the page
- *Black-out*, to "black-out" an area on the image that should be kept confidential
- *White-out*, similar to "black-out", except the area is *overlaid* with a white block
- *Inline Text*, to create text that can be positioned anywhere on the image

Implementation Approach

NYC and Xerox/USI used a standard methodology for deploying all Documetrix Document Management Systems. It focused on customer intervention and feedback as the principle axiom. As Business Process Analysts and Integrators, Xerox/USI acts as a facilitator by collecting the requirements and desires of the users and developing alternative processes for enhancing worker productivity and efficiency.

The Requirements Verification and Process Reengineering Tasks were utilized to analyze and evaluate critical processes to enhance their productivity. The tasks associated with each phase of the project are listed below.

Project Tasks:

- Orientation/Kickoff
- Requirements Verification
- Process Reengineering
- Technical Design Specification
- System Development
- Site Preparation
- User Testing and Acceptance
- System Installation (includes: Acceptance Test, Documentation, Training)

The following discusses some of the details of the implementation process and methodologies.

Requirements Verification

The principal objective of the Requirements Verification task was to confirm the information contained in the RFP and determine whether any unidentified requirements or assumptions exist, and to resolve any open issues. This was a formal process during which USI studies the client's operations, procedures, and infrastructure. An important component of this process was interviews with representative client staff.

During the requirements phase, USI personnel observed current procedures and discussed current and new procedures with the end-user and technical staffs. The Requirements Verification process helped USI gain an indepth understanding in areas such as indexing requirements (fields, sizes, and types), document relationships, user familiarity with GUI specific screen layouts, mainframe usage and processing, and other system interface requirements.

Upon completion of the Requirements Verification task, USI conducted a working session to review system requirements and objectives, provide input to the system design, and identify system constraints. The objectives of this session were to:

- Review proposal requirements;
- Identify additional requirements;
- Review system processes;
- Gain concurrence on an initial Logical Model; and
- Provide input into the Detailed System Design Specification.

System processes were discussed, relevant procedural documentation reviewed, sample documents collected, and workload estimates verified. This information provided USI with an in-depth understanding of the operational environment to design and develop the system accurately to meet NYC's needs.

The Requirements Verification Task is closely related to the Process Reengineering Task and provided it with pertinent process information. Some of the activities of these two tasks were performed in parallel.

Process Reengineering/Workflow Analysis

NYC recognized that the deployment of imaging, as an investment in technology, required the detailed analysis of the organization's business processes and current workflows. USI performed a high level process Reengineering effort for NYC.

USI system success in deploying new technologies is directly related to our ability to redesign our client's organizational workflow and the subsequent application of technology to achieve strategic objectives. The process utilized by USI is reflected in the following outline:

Workflow Implementation Process

- Current Processes
- Benchmarking/Best Practices
- Organization and Culture Assessment
- Technology Availability

- Visioning
- Redesign
- Prototype Review/Testing
- Cost/Benefit/Risk/Analysis
- Migration Planning

Detailed Design

Specify
- Performance
- Output
- Structure
- Organization
- Culture

Current Flows vs. Reengineered
- Recognized Benefits
- Human Performance Gains

Transition
- Systems Content
- Short-term Opportunities

Process/System Building
- Empowerment/Process Preparation
- Organization Preparation
- Training Development
- Progress Monitoring
- Acquire Software/Technology
- System Construction

The detailed work processes developed provided the foundation for the redesign efforts during the Process Reengineering Workflow Analysis Task. Additionally, the detailed work processes provided NYC a measurement for the validation of the cost/benefit analysis performed for this application.

Reengineering:
- Eliminated/Streamlined tasks
- Eliminated bottlenecks and delays between steps
- Enabled work to be processed in parallel rather than serially
- Provided simultaneous access to documents by multiple departments/people
- Provided greater control and security over the documents
- Allowed for quick, simple access to information, and
- Eliminated rework/retyping, while providing broader responsibilities

Resulting in:
- Improved productivity
- Reduced cycle time to complete work
- Reduced costs

- Improved customer service and public access, and
- Improved quality, accuracy, and consistency of results.

Technical Design Specification

Information collected during the Requirements Verification and Process Reengineering Tasks were used to develop the Technical Design Specification (TDS). The TDS included the following:

- Final version of the imaging hardware configuration
- Final version of the Detailed System Design (including the Database Design, Data Dictionary, Sample Screen Layouts, Menu Hierarchy, Information Flows)
- Project Completion Schedule (approved by the USI/NYC technical team)

The first step in the Technical Design Specification Phase was to gain formal acceptance and approval of the requirements and reengineered process specifications captured during the previous tasks. This acceptance was key to developing a successful system. Based on the original design concept developed in the Logical Model, USI developed a detailed Technical Design Specification that validated the system hardware and software architecture and system configuration for each Phase. This specification documented the application system inputs/outputs and system interfaces, database design (Entity/Relationship diagram), data dictionary, screen designs, indices, access methods, procedures, and processes, and report formats. Illustrations were provided detailing the hardware configuration, software components, and system links.

System Development

Upon acceptance and approval of the Technical Design Specification, USI began the System Development process. USI adopted NYC programmers as part of the development team. Assignments were given to NYC programmers in all aspects of Documetrix Document Management/Workflow development. USI's goal was for the NYC programmers to be self-sufficient in using the Documetrix product once the project was complete.

During the software development period, many prototype reviews were conducted at NYC to allow end users feedback on the system while it was developed. This hands-on interaction gave end users and managers an opportunity to become familiar with the system, and ensured that the functionality of the system clearly addressed NYC's requirements.

Change Management Process

OAISIS presented a major cultural change. This is due to the fact that for the last 100 years, NYC operated under a paper-based data-distribution process.

To ease this transition period, many initiatives were put into place. These included:

- Involving middle-management and subordinates in the prototype reviews
- Providing forums for questions and concerns of NYC employees
- Meeting early on with the NYC Labor representatives to give an overview of the project to get their *buy-in*

As the implementation date drew near, NYC published newsletters and Xerox/USI/NYC provided extensive training on OAISIS functionality.

Level of Overall System Complexity

Implementation of OAISIS was very complex for a variety of reasons, such as:

- Intensive Requirements Analysis
- Integration of NYC developmental staff
- Volume requirements of claim/contract paper documents
- Remote site network connections
- Conversion of millions of records from the CIMS/CATS mainframe systems
- Interaction with other mainframe systems
- Graphical user interface that provided an intuitive look and feel

Requirements Analysis

Requirements analysis for OAISIS started in August 1992. The claims application came on-line in August 1994. Xerox/USI along with NYC spent almost two years performing requirements, design, development, and installation.

Integration of NYC developmental staff

NYC developers were engaged as skills-transfer staff. One of the goals was for NYC to learn how to maintain, enhance, and develop new imaging applications. The NYC developers chosen as part of this task came from a mainframe background that presented a challenge to the Xerox/USI team to educate them in client-server developmental techniques such as SQL.

Volume requirements of claim/contract paper documents

The system needed to be sized correctly to handle the tremendous volume of paper that comes to NYC per year. NYC receives approximately nine million images per year, equating to about 36,000 images per day.

The massive volume of images proved a challenging effort in the development of the input sub-system along with storage of these images. It is a tribute to NYC and Xerox/USI that there is currently only a 24-hour backlog for daily, scanned images.

Remote site network connections

OAISIS is accessible to many NYC locations. Most of the locations were in Manhattan. However, there were other locations in Staten Island, Brooklyn, and Queens.

The integration of NYC's communication resources (DOITT, Nynex, and BIS technology) along with Xerox/USI WAN experience proved a challenging network project. Today, all these locations are part of the Document Management System accessing the same optical resources. Most optical queries result in image display in under five seconds. The performance of the OAISIS network, alleviates the old method of hand-delivering documents to these sites.

Conversion of millions of records from the CIMS/CATS mainframe systems

The following were some of the steps/issues that were addressed when Xerox/USI converted the mainframe database.

Mainframe Operations

- Discussed all mainframe preparations, to include:
- Storage space required

- Listing all jobs that interact with the mainframe system
- T-1 line preparation-the transference of data to the LAN

Extraction of Mainframe Data

- Extract format (comma delimited, fixed length)
- Print the test scripts used to verify each extract
- Approximate storage space per extract
- Approximate time to export per extract
- Person/department responsible for running the programs
- Reports that were (or could be) developed to test the validity of data when imported into Oracle (such as the number of cases, sum all cases settlement amounts, number of cases per insurance company, etc.)

Transfer Mainframe Extracts to NYC LAN

- Allocate enough storage space on the Novell File Server for the mainframe extract files
- Allocate enough storage space on the RS6000 for the mainframe extract files

Create Oracle tables

- Identify the Oracle database size
- Print the control files per extract file
- Print the Oracle table creation scripts
- Print the Oracle index creation scripts
- Run the table creation/index scripts
- Load the extracts into Oracle using a Loader utility
- Verify data has been loaded correctly (log files)
- When validated, export the tables. Will be stored away for future/historical use.

Issues that were addressed

- Translation Issues, such as:
- disposition codes, user identities
- Audit trails and history items that may be *archived*, and need to be reloaded
- Cleanup Issues
- Creation of statistical reports that previously operated on the mainframe.

Interaction with other mainframe systems

Both the claims and contract systems have been integrated with other NYC mainframe systems.

For example, OAISIS generates all data associated with the payout of claims. However, NYC has a warrant system (IFMS) that produces the checks. OAISIS has an automated job transfer server that sends vouchers to IFMS daily. The job server also receives information from IFMS that updates OAISIS with information such as warrant creation date. OAISIS averages over one million dollars daily.

Another example is the contract information comes from an NYC mainframe system called the Integrated Comprehensive Contract Information System (ICCIS). Each night information is downloaded from ICCIS to OAISIS. This greatly aids the Contract Indexing process.

Other examples of integration include Integration with a Mainframe System for the:

- Parking Bureau
- Office of Management and Budget
- Human Resources Department
- Law Department
- Building Information System
- Department of Environment Protection

Graphical user interface

OAISIS was tailored specifically to the needs of a large claims and contracts processing environment, with intuitive 'Tab' motif screens that resemble the file folders that users can easily relate to. Items that have been routed through the Workflow to a particular user are placed in an electronic basket. A variety of queue types can be chosen, such as a Manual Fetch Queue, in which users pull items from a public area; or Automated Queues which route documents to specific workstations based on decisions which the system can make automatically

Level of Integration with Other Advanced Technologies

OAISIS integrated with many technologies. They include:

- The scan workstations are integrated with a separate SCSI hardware component that provides an automatic page detection mechanism which alleviates the scan operators from manually setting page sizes.
- Barcode technology is used to further automate the indexing process.
- A graphical outline control was developed to provide an easy graphical method for indexers to manage batches of images.
- Documetrix workflow was used to electronically route claim/contracts to system users.
- Automatic printing of Boilerplate Letters were integrated with OAISIS.
- A Workflow device server was integrated.
- Transfer of data between OAISIS (client-server) to many NYC mainframe systems was delivered. Rumba was the mainframe emulation software integrated.
- Fax out technology was integrated.
- Fax in technology was integrated. This basically removed the scan step.
- Softscan technology was integrated. Softscanning is the process of turning a document, such as a Microsoft Word document into an image. This image would subsequently be stored on optical and provide a permanent, unalterable record of the document. Integration of this technology removed the step of printing and then scanning.
- Two 12" Sony Jukeboxes were integrated.
- Two Tricord Enterprise Novell File Servers were integrated.
- An IBM RS6000 database server was integrated.
- A wide-area network throughout the NYC area was implemented.
- A photograph scanner using JPEG compression was implemented.

- An engineering scanner was implemented using a compression format of Group IV Tiled. These are for paper that are up to E-size.

Scope and Scale of the Implementation

OAISIS is designed to provide the City of New York with an Electronic Document Management System, utilizing document imaging and workflow processing technology to support contract registration and claims processing functions.

The OAISIS scope included the following:

- Provided a Document Management Solution
- Replaced the paper-based claim files with image-based files
- Replaced the paper-based registration and storage of City contracts
- Replaced the Claim Information Management System (CIMS) mainframe application with a client-server database that includes CIMS functionality with the indexing and image management required for OAISIS
- Replaced the Contract Administration and Tracking System (CATS) mainframe application with a database that includes CATS functionality with the indexing and image management required for OAISIS
- Automated the workflow for claims investigation, approval, payment, and management
- Automated the workflow for contract registration and registration management
- Reengineered the entire claims and contracts processing activity
- A Wide Area Network that allows the New York City Borough offices access to the Optical System
- Incorporation of The NYC Comptroller's MIS staff (primarily mainframe backgrounds) as part of the USI/Xerox Development Team.
- Interoperability with mainframe and client-server based systems.

There are over 200 users of the system throughout New York City, primarily consisting of users from The Bureau of Law and Adjustment (BLA), and The Office of Contract Administration (OCA).

Impact

Extent and Impact of Demonstrated Productivity Improvements

The New York City Office of the Comptroller has realized extensive productivity gains using OAISIS. The number of new claim filings increased to record levels (The City is averaging over 30,000 new claims annually), while the number of employees has declined significantly. In 1990, before the system was implemented the Bureau responsible for claim processing had 142 employees. It currently has 93. This represents a total decrease in staff of 34 percent. During the same period, new claims have risen 13 percent over the same time.

Despite the severe decrease in staffing and the increased workload, since the system has gone into production, the active caseload has roughly halved. There are many reasons for this. OAISIS is a powerful tool that NYC employees can use to efficiently complete their investigations, but what really helped was the system's ability to process many work items that previously had to be done by hand. For example, OAISIS automatically closes out cases that have reached a particular statutory time limit. This used to be an extensive clerical task which always was behind and erroneously inflated current backlogs.

Immediately after OAISIS implementation, management was able to redeploy ten employees who earned a total of $300,000 annually to other areas of the Comptroller's Office. This worked out extremely well for the agency because it was a way to replace employees who had left City service because of early retirement incentives without productivity shortfalls.

But this is only part of the story. The system also has allowed management to use existing staff in better ways that have helped both to enhance revenues and control claim costs. For example, NYC was never able to find the resources in their paper-oriented operation to staff a unit totally dedicated to detecting and uncovering claim fraud. Not only have productivity improvements allowed management to create a position to coordinate fraud investigation operation, but the system has allowed them to uncover patterns of fraud and compare documentation in a way that they could never do using paper files. Since this one-person unit was created, the comptroller's office has identified over $200,000 in claims that it did not need to pay.

In another instance, for the first time, the Comptroller's office was able to create a unit to develop claims against other parties that damaged City property. Instead of this office paying out claims only for damage for which the City may have been responsible, the City now has a source of revenue collection. Since the effort started, over $180,000 in revenue has been collected.

The claims operation intention was always to settle personal injury claims at an early stage, however, resource shortages prevented management from implementing a program for early settlement. Analysis shows that if claims are settled before litigation starts, the average claim settlement is approximately halved. Using OAISIS, the Comptroller was able to establish a pilot project where nearly 150 claims were settled at an early stage. Over $1 million in settlement costs alone has been saved (this does not include further investigation and litigation expenses had the claims eventually gone into litigation). Based on the success of the pilot program, management plans to expand efforts in this area significantly.

The system performs many functions that were previously labor intensive and really subtracted from the quality and efficiency of investigations. Claims are now screened through workflow to determine if they have been filed in accordance with statutory requirements. Acknowledgment letters are generated automatically. Claims are assigned and routed to specific work locations automatically with little or no resource allocation. Status letters are automatically generated by the system for certain claim types, thus allowing the Comptroller's office to keep claimants informed of their claims two months, five months and one year from the date of their filing the claims. All this is done automatically by the system.

The system allows multiple users to process different aspects of the claim at the same time. For example, an electronic claim file is routed to the hearing division for hearing scheduling at the same time it is routed to a claim examiner for investigation. Both tasks can be accomplished simultaneously without the physical transfer of a paper file.

Workflow tools allow management the ability to schedule individual systematic claim review without a disruption to the investigation. Management can also see the entire claim process graphically and determine bottlenecks. Thus, additional resources can be deployed where they are needed without a management analysis of a particular process problem.

Workflow allows management to modify business processes in real time. Users can streamline processes by essentially drawing a graphical line between two points on existing workflow models. Additionally, individual files can be transferred on line from one employee to another within seconds. This enables management to resolve immediately problems that arise in the course of business when an employee is out for extended periods or is reassigned to other tasks.

Claim files are now valuable investigative tools. There are no duplicate copies of documents and each document is indexed and sorted into a particular location in the electronic file. Users have access to "libraries" which contain incident reports that are systematically collected from other city agencies. These reports can be placed in individual claim files as appropriate.

The Comptroller's office has access to a number of city and state agency computer systems. The staff is able to automatically transfer images of these systems' data to a particular part of the claim file for permanent record. This has saved countless hours in additional processing time. Users have the ability to create documents in a word-processing application and, through a process known as "softscan," add the document automatically to the claim file and simultaneously print a copy out for mailing.

As a result of the nature of the system backup processes, the City will be able to withstand damage caused by fire or flood to claim files. Data that appears in files is secured in ways that are not available using paper files.

The staff's ability to respond to telephone inquiries, which represents a substantial part of a claim examiner's day has been greatly enhanced. A claim examiner can retrieve virtually any claim file within five seconds and is able to answer callers without an unreasonable delay or call back.

OAISIS also captures images of all city-registered contracts. claim examiners can access and use contract information to investigate claims, specifically certain claims may involve losses due to a contractor's performance on a particular city job. The contract can be used to identify the availability of contractor insurance to indemnify, exonerate or defend the city.

Significant Cost Savings

Overall the City estimates that it will either save or generate revenues of over $20 million over the next five years. OAISIS has already:

- Saved $300,000 in direct labor costs annually by redeploying 10 employees.
- Identified over $200,000 in fraudulent claims.
- Saved over $2 million in settlement costs by settling claims early. This initiative was supported exclusively by OAISIS.
- Saved over $1 million annually in mainframe system support costs.
- Saved $130,000 in storage and reproduction costs during the first year of operation.
- Collected over $250,000 in affirmative claims.

Culture Change

Before the City implemented OAISIS, they operated in virtually the same way as they operated prior to and during the twentieth century. They relied mainly on manual processes to deal with an ever-increasing workload. Before the system was put into production, only one personnel computer was assigned to the claims area. They were supported by a mainframe system that was not interactive and basically maintained data that was used to calculate the City's long term liability but gave little support to the claim investigation and settlement process.

The Bureau's files were located in approximately seven locations throughout the City. Much of their clerical resources were spent keeping track of and locating files. Only one person was able to work on a file at a time. When litigation started, the Bureau lost control over its own files, which were transferred to the City's attorneys. Three people were devoted exclusively to locating and preparing files to go to the Law Department.

The Bureau was running out of room to store claim files and they were susceptible to fire or water destruction. The lack of adequate file space caused the Bureau to keep them boxed in hallways and work areas. In fact, the file storage conditions made Bureau offices the backdrop for a number of television and film screenings *(Law and Order* and *Q&A*). Certainly it was not a great environment for employees to maximize production.

Most of the Bureau's efforts were devoted to moving paper files from one office location to another. There were no resources available to conduct fraud investigations, early settlement programs and affirmative claim actions. The ability to conduct a thorough and comprehensive claim examination was limited. The claim files themselves were replete with duplicate copies of every document and filed in a standard first in, last out order.

OAISIS has empowered employees. They are no longer exclusively concerned with maintaining the paper file. They are now looking at claim files with a view toward settlement at an early stage, as well as conducting more in-depth and comprehensive investigations. Claim examiners were basically performing clerical functions before OAISIS was implemented. Today they have the tools to conduct a professional claim investigation.

Question 1 Describe the system application. What is the system used for, who are the users and what does the job entail? How often or how many hours is the system in use on a daily basis?

The Omnibus Automated Image Storage and Information System (OAISIS) is designed to provide the City of New York with an Electronic Document Management System, utilizing document imaging and workflow processing technology to support contract registration and claims processing functions.

To summarize briefly, the OAISIS project entailed the following:

- Provided a Document Management Solution
- Replaced paper-based claim files with image-based files
- Replaced paper-based registration and storage of City contracts
- Replaced the Claim Information Management System (CIMS) mainframe application with a client-server database that includes CIMS functionality with the indexing and image management required for OAISIS

- Replaced the Contract Administration and Tracking System (CATS) mainframe application with a database that includes CATS functionality with the indexing and image management required for OAISIS
- Automated the workflow for claims investigation, approval, payment, and management
- Automated the workflow for contract registration and registration management
- Reengineered the entire claims and contracts processing activity
- Integrated a Wide Area Network that allows the New York City Borough offices access to the Optical System
- Incorporated the NYC Comptroller's MIS staff (primarily mainframe backgrounds) as part of the USI/Xerox Development Team.
- Integrated mainframe and client-server based systems.

There are over 200 users of the system throughout New York City, primarily consisting of users from the Bureau of Law and Adjustment (BLA), and the Office of Contract Administration (OCA).

The system is in use from 7:00am-midnight. A second shift of City users working in the Central Imaging Facility (CIF) makes up the majority of the later hours.

Question 2. What were the key motivations behind installing this system?

The New York City Office of the Comptroller's Bureau of Law and Adjustment had been operating in an early twentieth-century business environment. Up until 1975, the office did not use computers for processing any claim work. Even when the first mainframe claim database (Claims Information System) was installed in 1975-6, its use was confined to keying in basic claims information. This was a duplication of work being done on the paper file itself. The improved system that followed the Claim Information System was another mainframe application known as the Claims Information Management System. This system involved more user input and increased the integrity of the claim numbering system. It also allowed for the monthly production of claim productivity statistics and form letters by division. It, too, required duplicating work which had already gone into the paper file.

The New York City Office of the Comptroller had a need for an integrated computer system that would allow the staff the ability to perform all their work at one terminal. The creation of software that permitted a multi-tasking work environment, the improvement in scanning technology and the use of workflow software all combined to enable the City to envision a new work standard. Combined, these technologies transformed the work environment and the product produced. The City had to find ways to perform its mission more efficiently and effectively, which translates to reducing upfront costs in the creation and investigation of claims to the final resolution and payment of claims at the best possible cost, involving the least number of work steps and staff necessary to accomplish the task.

With this scenario in place, the City could concentrate some effort in reducing its liability by analyzing the claim base for fraudulent claims. It instituted early settlement processes which traditionally involved less cost in settlement payouts, provided extracts of the database to other City agencies with collection authority to allow data comparisons that would identify claimants who owed the city money for one reason or another.

The City could also aggressively seek affirmative claim situations whereby the City could recoup moneys when other parties were found liable for damage to city property.

Question 3. Please describe the current system configuration (number and type of software, servers, scanners, printers, storage devices, etc.)

NYC and Xerox/USI conducted a comprehensive review of the entire process of claims and contracts within the Comptroller's office. The new system needed to handle a multitude of documents in a range of sizes. In addition to meeting the current requirements, the system needed to interface with other city organizations, as well as various computer and database systems that were already in place. OAISIS needed to be a comprehensive image and workflow management system, tailored specifically to the needs of the Comptroller's office and designed around the established information infrastructure.

NYC worked with Xerox/USI's engineers to make significant enhancements to the network in place so that it would be able to handle the tremendous increase in bandwidth that an image processing system demands. The OAISIS WAN spans over five sites, with well over 200 users. Two Tricord Novell file servers are employed.

Software Used

Documetrix software was used to develop the Document Management System. The Documetrix solution is built upon an open systems environment client-server architecture. The open systems approach enables the support of a variety of databases (e.g., Sybase, Oracle, Informix, Ingres, etc.), a variety of networking environments (e.g., Novell NetWare, TCP/IP, Banyan VINES, Microsoft, LAN Manager, NT) or operating environments (e.g., UNIX, NT, Novell). The client-server architecture provides NYC with the modularity, scalability and flexibility that the client-server solutions can offer.

Documetrix is scaleable because it is based on a client-server architecture. Client-server architecture spreads functionality across separate servers to optimize the workload for the system. The open client-server architecture of the Documetrix solution enabled Xerox/USI to provide NYC with a solution that provides the core functionality as more applications are developed and come on-line. The following are the Documetrix software utilized.

Documetrix 2000 Server
Documetrix Desktop Client
Documetrix Workflow
Documetrix Workflow Client
Documetrix AutoIndex Server
Documetrix 2000 API
Documetrix Workflow API
Documetrix Scan Software
Documetrix/Teamworks Print Server

OAISIS was tailored specifically to the needs of a large claim and contracts processing environment, with intuitive 'Tab' motif screens that resemble the file folders that users can easily relate to. Items that have been routed through the Workflow to a particular user are placed in an electronic basket. Any of a variety of queue types can be chosen, such as a Manual Fetch Queue, in which users pull items from a public area; or Automated Queues which route documents to specific workstations based on decisions which the system can make automatically. A series of device servers are present throughout the workflow which perform automatic processes such as printing acknowledgment letters and automatic removal of disallowed claims.

File Server Used

OAISIS employed the Tricord PowerFrame Enterprise server. This server has a 64-bit internal system bus that is transparent to all operating systems and provides a bandwidth of 267 Mbytes/second. This server is running Novell NetWare.

The Enterprise Server comes with ten PowerBus slots that are used to accommodate a dedicated MMS (Main Memory SubSystem), IMS (Intelligent Management SubSystem) processor, a slot is also used to accommodate the ISS (Intelligent Storage SubSystem) and a CCS (CPU/Cache SubSystem) leaving slots for expansion.

The SYS volume of the "Tricord" server is duplexed using two 3GB drives. The SCAN and CACHE volumes are configured on a RAID 5-disk array covering 9 of the remaining 3GB drives. The SCAN volume is 10GB, the CACHE volume is 11GB. The remaining two disk drives are configured as on-line hot spares which automatically get used in the event of failure of any of the other drives in the server.

The CCS provides high-performance multiprocessing. The MMS supports up to 1 gigabyte of 128bit ECC DRAM (most PC based servers utilize 64bit) can be installed on the MMS. All of the components on the PowerBus can access this memory at 267 Mbytes/second. The OAISIS file server is configured with 254 Mbytes of RAM.

The ISS (Intelligent Storage SubSystem) is Tricord's SCSI controller that dedicates an Intel microprocessor to handle disk I/O tasks. By utilizing a dedicated processor, the ISS typically offloads 90 percent of disk I/O processing which in PC based server platforms the CPU itself must handle. The OAISIS file server is configured with over 21 gigabytes of hard disk. With approximately 11 gigabytes used to store temporary scanned images, and eight gigabytes used for an optical cache.

Database Server Used

The OAISIS contract and claims applications use an IBM RS/6000 as the database server. The advanced RISC architecture allows the RS/6000 to support the OAISIS contract and claims applications. RISC computers feature low-level instructions, all of which are executed in hardware. For an RS/6000, the RISC hardware takes the form of a proprietary set of three chips. This RISC implementation combined with wide bus paths, high clock speeds, and large cache memories, provide OAISIS superior database performance.

The RS/6000 runs the AIX multi-user operating system. AIX provides the benefits associated with UNIX, including standardization that eliminates worries about continued support for the product by IBM.

The architecture for the RS/6000 is the 580 model with 14 GB of RAID5 magnetic disk in two Data General Clarion disk subsystems. The disk size is based on Xerox estimates and experience and based on the complexity and size of the database. This is an estimate for three years' processing based on a table space of 12K per file for claims and contracts as well as an additional 2K per claim page and 20K per contract file for annotations. The 12K includes index as well as workflow data.

The unit is configured with an FDDI network interface card for attachment to the OAISIS FDDI ring.

Scanners

The scan capabilities are extensive, with four main Bell and Howell 6338 top-of-the-line high speed scanners acting as primary source of document conversion, as well as several off-site scanners in various department including Legal, Engineering and the Hearing Department. High-throughput scanners are needed given that the average daily volume is 36,000 images per day. The scan workstations are configured with an automatic page detection mechanism that alleviates the scan operators from manually setting page sizes. In addition, bar-code technology is used to further automate the indexing process.

After the documents are indexed, they are injected into workflow. OAISIS uses Documetrix Workflow to route claims and contracts electronically. Workflow allows documents to be routed and tracked across the Wide Area Network to various buildings throughout the city via several high volume T1 connections. An IBM RS-6000 acts as the database server, housing the Oracle RDBMS. Document archival is handled using one of two Sony Optical Jukeboxes, one for claims and the other for contracts.

Storage Devices

The purpose of an optical subsystem is to handle requests for images which are passed to it from the application. The optical subsystem is comprised of two SCSI interface cards installed in the Tricord server. Each of these SCSI controllers is attached to a Sony 930 jukebox. The subsystem is controlled by a series of NLMs which are loaded on the Tricord.

The subsystem receives and queues requests for data stored on the optical jukebox requested by OAISIS workstations..

Write Once Read Many (WORM) technology is used to provide permanent archival storage. The Documetrix software running on the server, controlling the optical subsystem is designed specifically for a multi-user imaging environment. Specific caching and addressing schemes have been developed to optimize throughput. In addition, as the OAISIS requirements expand and the technology continues to evolve, new storage devices and media types are easily incorporated into the Documetrix platform via the incremental addition of document servers.

The jukeboxes are configured to support 12" WORM optical media. This size of disk was chosen because of the resulting capacity, performance and management characteristics. A 12" platter holds 6.4GB of data, 3.2 on each size as opposed to a 5.25" disk that holds 1.4GB, 700 MB on each side. The selected jukeboxes also have a faster disk-swapping mechanism, spin-up time as well as higher throughput than smaller media jukeboxes.

The claim jukebox is configured with three 12" optical disk drives and a storage capacity of 57 platters. The contract jukebox is configured with two drives with a capacity of 67 platters. The difference in jukebox configuration is a reflection of the differences in requirements of image data retrieval and storage between OCA and BLA. Contract's optical requirements are based on storage requirements rather than retrieval. Claims is based on retrieval first, storage second.

Workflow Server

The workflow server is the set of programs that "run" the workflow. The workflow server populates and maintains the workflow in-progress tables and the workflow history tables. In addition, the workflow server is responsible for distributing work items to groups of users or devices.

Fax Server

The fax server has the function of receiving or sending fax images. This allows Documetrix users to choose a page of a document and have it faxed. For example, users may select an image or a form letter (such as a request for information from a claimant/attorney) and have it faxed automatically without having to leave their desks. Additionally, having documents faxed-in alleviates the need for scanning them into the system.

Barcode Server

The OCR barcode server receives data containing barcodes needing recognition. A function to be performed is encoded into the barcode which the OCR barcode server can recognize. The barcode server can then decode the barcode and perform the encoded function.

The OCR barcode server is attached to the network and is dedicated to recognizing barcodes. The Documetrix software running on the barcode server is designed to work with most common barcode formats, such as code 3 of 9 or code 128. Xerox has barcode server products which support either code and has chosen code 3 of 9 because it is becoming a de facto industry standard.

OAISIS index information is encoded into the barcode. The OCR barcode server recognizes the information encoded into the barcode. During the scanning process a barcode sheet can be scanned along with contract and claim files. The barcode contains contract or claim document divider information. The barcode server then begins the process of indexing this contract or claim information into OAISIS. This reduces the amount of indexing information needed to be data entered during the indexing process.

Printers

Most print servers installed for OAISIS are those connected to a standard HP LaserJet 4 printer. A special board is required to allow the LaserJet 4 to print images at its rated speed of eight pages per minute. This board is a image decompression board which is installed in the expansion slot of the printer. The image decompression board has a standard parallel port on it which allows for a variety of connection methods for the printer to be accessible to OAISIS users. OAISIS has over two dozen print servers.

Question 4. How is this system integrated with the company's other information processing systems?

OAISIS had to be integrated with the existing infrastructure of NYC, to include not only the components on the Wide Area Network (WAN) but also other NYC mainframe applications.

WAN integration

The OAISIS WAN has been designed to support all users of the OAISIS imaging system using the most cost effective, highest performance components available in the market.

The Municipal Building at 1 Centre Street in Manhattan serves as the central hub for the OAISIS WAN. Communication lines from the Municipal Building to the satellite locations will service the connections to the imaging system from each of the remote locations.

The selection of communications link speeds and media is based on cost, performance and capabilities. The number of workstations at the remote location determines the speed and bandwidth required for acceptable response times on the imaging system. The preferred method of connection of all workstations is through the use of media that can support the highest speeds. For OAISIS, the preferred method of connection of remote sites is through a hard-wired (fiber) link that will support communications at Ethernet speeds (10 Mbps). Xerox will use this capability where it exists, between the Municipal Building and the Comptroller Engineering/Real Property offices at 2 Lafayette Street. Fiber cable has been installed between the buildings and sufficient strands are available for OAISIS use.

Barring the availability of existing communications lines capable of supporting the OAISIS protocols (IPX and IP), installation of one or more lines has taken place. Installation of fiber optic links between the Municipal Building and the remote locations would be the optimum method of connection of all sites. These lines can support high speed and bandwidth communications and are not subject to monthly payments or as susceptible to outages or performance problems. Unfortunately the installation of dedicated fiber links is extremely expensive and often logistically impossible because of the path the fiber must travel from point to point.

The most frequently used remote transmission technologies include dedicated T-1 and fractional T-1 lines. These lines are available in increments of 56 Kbps up to T-1 speed which uses a speed of 1.544 Mbps. These lines are available from NYNEX through CityNet. The initiation of such a line requires a one time installation charge and a recurring lease cost for monthly usage of the line. The use of dedicated leased lines provide a more stable means of data communications than dial-up lines

One large aspect of OAISIS was to use the existing network infrastructure as the basis for the Document Management System. To this end, OAISIS was built upon the existing network of NYC to create the WAN. The success of the OAISIS WAN integrated with the existing network can be attributed to the teamwork environment between Xerox and NYC.

Mainframe integration

Both the claims and contract systems have been integrated with other NYC and New York State mainframe systems.

For example, OAISIS generates all data associated with the payout of claims. However, NYC has a warrant system Integrated Financial Management System (IFMS) that produces the checks. OAISIS has an automated job transfer server that sends vouchers to IFMS daily. The job server also receives information from IFMS that updates OAISIS with information such as warrant creation date. OAISIS averages over one million daily.

Another example; the contract information comes from a NYC mainframe system called (ICCIS). Each night information is downloaded from ICCIS to OAISIS. This greatly aids the Contract Indexing process.

Question 5. Describe how the company has been impacted by this system. Be as specific as possible: What productivity improvements have been realized? How has the business workflow been affected (compared to before system implementation

OAISIS has made a dramatic impact at every level of processing claims in every division. It has transformed traditional workflow concepts into graphically mapped and logically linked tasks and flows. There are many templates that cover the entire spectrum of claims processing. Workflow allows for instantaneous modification of any divisions task or flow for maximum efficiency and effectiveness.

The system, for the first time, allows management the capability to measure and quantify staff productivity. It allows management the ability to maintain workflow and intercept bottlenecks before they become insurmountable problems. OAISIS provides a complete claim processing environment from the intake and coding of new claims to the disposition, settlement and payment of claims. In the event of a disaster, the entire system is backed up and can be restored to a disaster recovery site to allow the continuation of work.

Statistical analyses of the claims database have improved significantly with OAISIS. The system is open and modifications such as other new applications can be integrated into the work environment with minimal cost.

In sum, OAISIS has transformed the New York City Office of the Comptroller's Bureau of Law and Adjustment into a twenty-first century claims operation.

Question 5A. What cost savings of increased revenues have been realized since the system was first installed?

Immediately after OAISIS implementation ten employees who earned a total of $300,000 annually were redeployed to other areas of the Comptroller's Office. A unit was created to develop claims against other parties that damaged City property. A pilot project was implemented to claims at an early stage.

Question 6. Describe the implementation process and methodology, the project team and any change in management and business process reengineering issues addressed.

NYC and Xerox/USI used a standard methodology for deploying all Documetrix Document Management Systems. It focused on customer intervention and feedback as the principle axiom. As Business Process Analysts and Integrators, Xerox/USI acts as a facilitator by collecting the requirements and desires of the users and developing alternative processes for enhancing worker productivity and efficiency.

Requirements Verification and Process Reengineering Tasks were utilized to analyze and evaluate critical processes to enhance their productivity. The tasks associated with each phase of the project are listed below.

Project Tasks:

- Orientation/Kickoff
- Requirements Verification
- Process Reengineering
- Technical Design Specification
- System Development
- Site Preparation
- User Testing and Acceptance
- System Installation (includes: Acceptance Test, Documentation, Training)

The following discusses details of the implementation process and methodologies:

Requirements Verification

The principal objective of the Requirements Verification task was to confirm the information contained in the RFP and determine if any unidentified requirements or assumptions exist, and to resolve any open issues. This was a formal process during which USI studied the clients' operations, procedures, and infrastructure. An important component was interviews with representative client staff.

During the requirements phase, USI personnel observed current procedures and discussed current and new procedures with the end-user and technical staffs. The Requirements Verification process helped USI gain an in-depth understanding in areas such as: indexing requirements (fields, sizes, and types), document relationships, user familiarity with GUI specific screen layouts, mainframe usage and processing, and other system interface requirements.

Upon completion of Requirements Verification task, USI conducted a working session to review system requirements and objectives, provide input to the system design, and identify system constraints. The objectives of this session were to:

- Review proposal requirements;
- Identify additional requirements;
- Review system processes;
- Gain concurrence on an initial Logical Model; and
- Provide input into the Detailed System Design Specification.

The Requirements Verification Task is closely related to the Process Reengineering Task and provided it with pertinent process information. Some of the activities of these two tasks were performed in parallel.

Process Reengineering/Workflow Analysis

NYC has recognized that the deployment of imaging, as an investment in technology, required the detailed analysis of the organization's business processes and current workflows. USI performed a high-level process reengineering effort for NYC.

The process utilized by USI is reflected in the following outline:

Workflow Implementation Process

- Current Processes
- Benchmarking/Best Practices
- Organization and Culture Assessment
- Technology Availability
- Visioning
- Redesign
- Prototype Review/Testing
- Cost/Benefit/Risk/Analysis
- Migration Planning

Detailed Design

Specify

- Performance
- Output
- Structure

- Organization
- Culture
- Current Flows vs. Reengineered
  Recognized Benefits
- Human Performance Gains
- Transition
- Systems Content
- Short-term Opportunities
  Process/System Building
- Empowerment/Process Preparation
- Organization Preparation
- Training Development
- Progress Monitoring
- Acquire Software/Technology
- System Construction

The detailed work processes developed provided the foundation for the redesign efforts during the Process Reengineering/Workflow Analysis Task. Additionally, the detailed work processes provided NYC a measurement for the validation of the cost/benefit analysis performed for this application.

The benefits of this reengineering included:

- Eliminated/Streamlined tasks
- Eliminated bottlenecks and delays between steps
- Enabled work to be processed in parallel rather than serially
- Provided simultaneous access to documents by multiple departments/people
- Provided greater control and security over the documents
- Allowed for quick, simple access to information, and
- Eliminated rework/retyping, while providing broader responsibilities

Resulting in:

- Improved productivity
- Reduced cycle time to complete work
- Reduced costs
- Improved customer service and public access, and
- Improved quality, accuracy, and consistency of results.

NEW YORK CITY OFFICE OF THE COMPTROLLER

OAISIS Diagrams

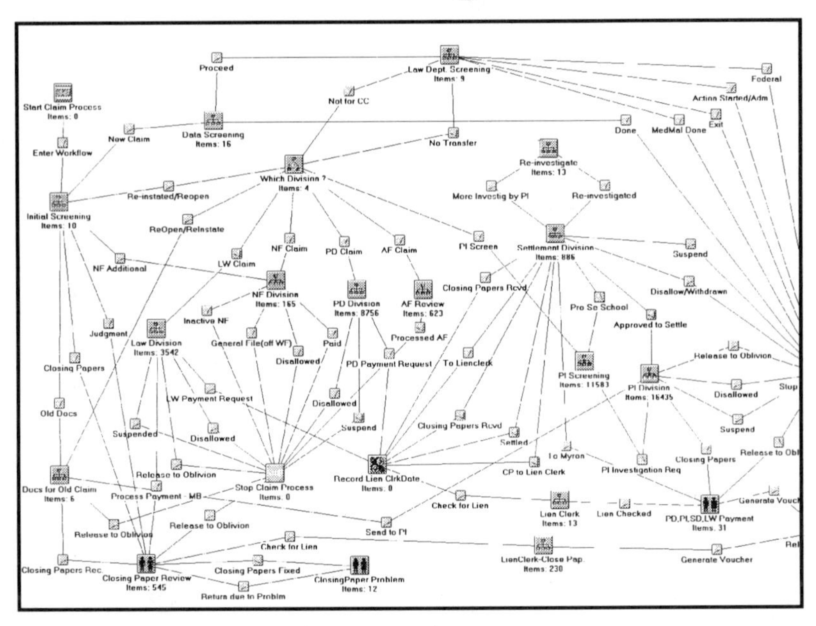

OAISIS Top Level Claim Workflow Template

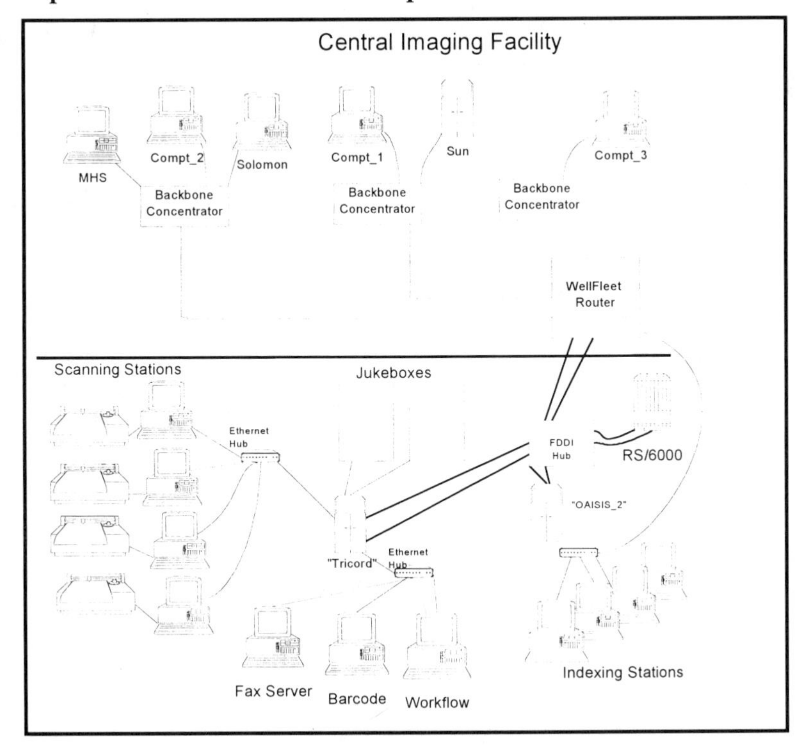

OAISIS Municipal Building Network located at 1 Centre Street

OAISIS Wide Area Network Locations of New York City

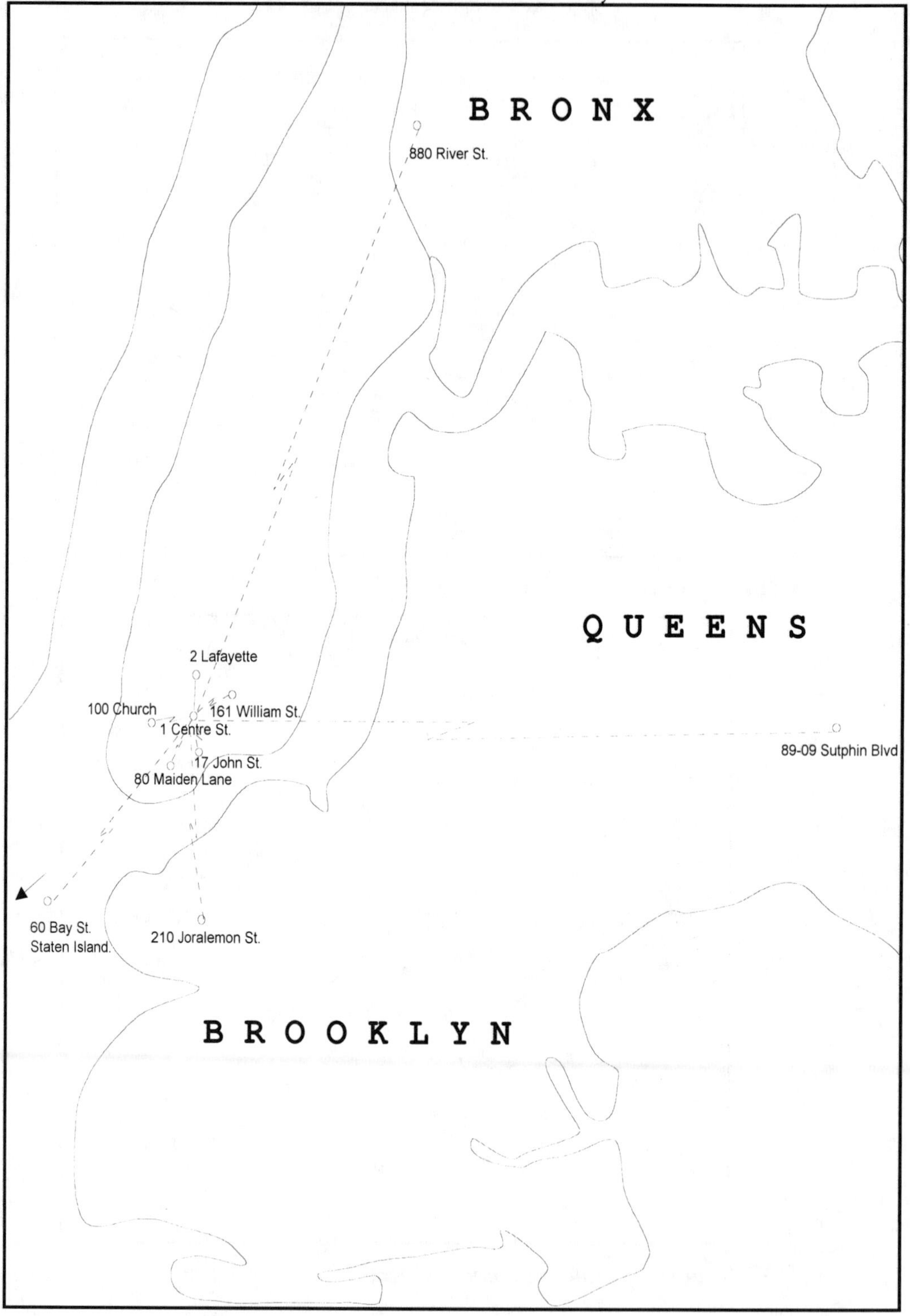

PPP Healthcare,
Tunbridge Wells, England

European Excellence Award: Workflow, Silver

Executive Summary

To unravel the mystery of business process automation, take 101 different document types, distribute around three LANs, one WAN and 640 desks. Tthrow in a jumble of hardware and software, superfast state-of-the-art scanners, jukeboxes and the latest in workflow technology, whiz together with a legacy mainframe system and it is possible to begin.

It is true that not every business process automation program will take this mix of new and used IT as its infrastructure. However, these are the raw ingredients exploited to great effect by England's PPP healthcare. PPP healthcare has implemented a workflow strategy, which is already delivering dividends by cutting out 'red tape' and increasing staff productivity. PPP healthcare's mission is to supply its 2.3 million customers with the ultimate in service. In fact, so successful has its implementation been, that it now claims to have an infrastructure that is increasing customer loyalty with visible improvements in service.

Customer Satisfaction is Key

PPP healthcare's brand awareness is very important to the Company and its future growth. It has steadfastly developed and activated high-profile marketing programs to promote its caring and personal approach, which it claims to provide every step of the way. To this end PPP healthcare has introduced a range of products and services to satisfy a myriad of healthcare needs, but products alone do not build a consistent market image.

Mike Tinsley, business project manager, planning and development division at PPP healthcare explained "We can all design and sell new products-but the reality is that people buy, and buy again, based on service, quality and satisfaction. It is the service in our business that will make or break customer loyalty."

This in some way explains why PPP healthcare has invested heavily in an IT infrastructure, which it expects to deliver on its brand promises.

"The technology we have in place today enables us to provide the best customer service in the business, in that we can respond to a customer inquiry instantly," he continued.

To reach the point whereby PPP healthcare can guarantee the best available service, it looked very closely at its customer support and response processes. At this time it recognized that workflow technology was key to overall improvements. PPP healthcare's process involves a large variety of inbound and outbound documents from various sources, such as general correspondence, claims, hospital and consultant/specialist provider accounts being merged into a single case file. Previously, this resulted in a lot of paper work, handling and filing. Following information collation and gathering, the file was passed to the responsible personal adviser for action. Once a file was closed, and a customer claim dealt with, it was archived on microfiche. The whole process was paper, people and time-intensive, with many opportunities for human error. Moreover, it required a large number of staff for the microfiche handling alone.

An Integrated Process

Having examined the process, PPP healthcare established that the new system needed to remove all opportunity for error and allow for future developments and growth. Consequently, the decision was taken to totally integrate the process from start to finish. The new system scans, indexes, merges legacy and current data on the fly, and delivers information automatically to the desktop of the designated personal adviser.

It goes a stage further than this; all staff has instant access to all information and any member of the team can provide a customer with exact, up-to-the-minute details on the status of his or her claim.

"This really is the beauty of the system, all documents are electronically available, managed, merged and produced out of the ViewStar workflow system which has been our route to integration on the desktop," Tinsley added.

Technology Requirements

PPP healthcare selected ViewStar as its workflow/business process environment, in late 1993 based on its open architecture, recognizing that total system integration on the desktop was imperative to the level of customer care and service PPP healthcare was determined to achieve. The whole system is designed to serve the needs of PPP Healthcare's three main offices in and around Tunbridge Wells, Kent and Eastbourne, Sussex. It comprises two Novell Local Area Networks connecting 16 NT Compaq servers and AST, WindowsTM clients.

There are two Kodak scanners on the third LAN taking care of incoming post and two Hewlett-Packard jukeboxes for optical disk storage and archiving. A Wide Area Network connects the LANs and users to the mainframe, which holds valuable customer legacy data. Currently scanning takes care of 10,000 documents per day, although when the system is running to full capacity, Tinsley expects to consume fully 30,000 documents into the workflow system daily.

"There are literally too many document types to mention, they could be anything from a birth certificate through to handwritten bills sent in by specialist providers-but whatever the content or type of document they will need to be assigned and managed," added Tinsley.

The ViewStar system integrates all components onto the desktop, merging, tracking and delivering information no matter its origin, for example, scanned documents or those generated internally by accounting or word processing applications. The workflow system also interacts with the mainframe, retrieving legacy data that resides on the Amdahl/MVS and here, ViewStar integrates with the terminal emulation software, Attachmate.

"The system represents the first true client server environment the Company has ever had; it required many changes in the IT department. We needed new skills for support and discovered that managing a client server environment is much more complex than managing a mainframe," said Tinsley.

Implementation

In the early stages of its implementation and in addition to the network issues, PPP healthcare encountered many unforeseen difficulties. For example, every time there was a technical problem, it had to be isolated so that the right supplier could be contacted for support. In addition to orchestrating the installation and system integration, PPP healthcare found that it had to overcome a massive learning curve.

"We became very frustrated, because the technology is much younger than, say the mainframe.. While PPP healthcare developed new personnel skills and recruited new staff, we quickly identified that the tools for managing the client server system and network lacked maturity in areas such as system performance monitoring and diagnostics," said Tinsley.

Having surmounted the technical problems and reconciled its IT skill base, PPP healthcare had yet another hurdle to jump.

"The biggest challenge came from a completely unexpected quarter. You see, we had previously taken telephone inquiries from customers and responded by calling them back. From our point of view, it is much more productive and service oriented to instantly access information while the caller is on the line. But we did not consider the 30 seconds the caller may have to wait while we retrieve the customer file. Thirty seconds is obviously a great improvement over, perhaps, half a day but when you are waiting on the end of a phone it can seem a long time," said Tinsley.

It is hard to imagine the frustration of PPP healthcare when it discovered that it had successfully set up and invested in a system that the customer perceived to be slow-even though it was actually delivering a much improved service. PPP healthcare understood that it had a weak link in its system. And the culprit was the highly visible personal computer. PPP healthcare's staff was equipped with 486-based PCs.

"By switching the 486s to Pentium-based technology we immediately improved the retrieval times— the result is the customer perceives instant response times," Tinsley explained.

Indeed, PPP healthcare's personal advisers wait no more than nine seconds to retrieve every relevant item of information. Besides upgrading its PCs, PPP Healthcare found that it could improve system performance by migrating its database to NT SQL Server.

The System's Impact

Today, ViewStar's workflow system is available to 300 PPP healthcare staff, with plans to further expand the system, making it the working environment for all of PPP healthcare's 640 staff who deal with customers by the middle of 1996. However, the Company is already benefiting from 10 percent gains in total productivity and is using the system as a marketing tool.

"We know that our recent promotional campaigns and revamped branding is educating the market and generating more business. However, the workflow system is proving an excellent selling point for corporate private healthcare buyers," said Tinsley.

Potential corporate customers are invariably shown a demonstration of the ViewStar business process automation system, thus proving PPP healthcare's commitment to service and clearly illustrating its ability to deliver against its brand promise.

By using ViewStar's open architecture to integrate the desktop across a sophisticated network environment, PPP healthcare can manage its cocktail of complex and varied documents within a single workflow solution. In doing so, PPP healthcare has unraveled the mystery of business process automation to benefit its customers, its staff and ultimately contribute to its future growth.

Red Ball Oxygen Company, Inc.
Shreveport, Louisiana

North America Excellence Awards: Imaging, Finalist

Executive Overview

Red Ball Oxygen employs a Minolta MI^{3}MS 3000 network with AOS and MI^{3}MS 3000 I.C.E. to help achieve one of its main business objectives: reduce *transaction costs,* that is, to reduce every step possible to minimize the time associated with any function involving employees. They discovered that imaging was one of the ways to accomplish this objective.

The situation and the problem

Red Ball distributes industrial gas cylinders to the welding supply industry and also, through a medical division, to hospitals, clinics and home users of oxygen in a territory of approximately a 150 mile radius from its headquarters The firm has 100 employees, seven branches and 5,000 active customers.

Accounting for the delivered cylinders is important not only because the cylinders represent a company asset, but also because of government regulations associated with the cylinder's content. A signed delivery receipt, therefore, is mandated, and must be kept on file. Approximately 120,000 deliveries are made in a year. A file is created for each delivery comprising the delivery receipt, customer invoice and attachments such as meter readings. According to Red Ball the volume of paper associated with each delivery was the foundation for their "file and misfile" system, as they called it, which overloaded the file room with five to six hundred thousand pages. To file these transactions, documents were first arranged numerically by delivery ticket/invoice number, then integrated into a temporary file, and then into a larger permanent file. Labor costs for filing, retrievals at a rate of ten to twenty per day and locating misfiles was estimated at $30,000.00 per year.

The solution

Working with Microfilm Supply Inc. of Monroe, Louisiana, Red Ball arrived at an image and information management system solution. A five-seat MI^{3}MS 3000 system was installed in August, 1995.

The application flows as follows:

- The drivers return signed delivery receipts at the end of the day. The documents may be preprinted on the distributor's AS/400 Advanced 36 mainframe computer or in some cases written by the drivers. The delivery ticket number becomes the subsequent invoice number.

- Data from the delivery receipts are entered to the AS/400, which then generates invoices.
- The invoices are downloaded daily to the imaging system and indexed through the MI3MS 3000 I.C.E. option (a COLD-type software module). Fields that are extracted include customer name, customer number and invoice number.
- Signed delivery receipts are scanned to a batch. To index the individual receipts, the indexing workstation operator enters only the invoice number. Through macros developed for this application by the customer, a number of functions occur automatically:
  - the remainder of the fields from the downloaded invoice are transferred to the imaging system index;
  - the image of the invoice is then displayed on the monitor side-by-side with the delivery receipt;
  - the image of the receipt is zoomed up to the receipt (invoice) number for verification;
  - when verification is complete, the receipt and attachments are saved on optical disk along with the invoice and the next item comes up automatically.

Red Ball has now stored a compound document consisting of a scanned, bit mapped image of the delivery receipt and attachments plus a character-based, computer-generated invoice. All of this has been accomplished with a minimum of manual indexing and zero toggling between or manipulation of different software modules.

By using the invoice file to create the bulk of the index along with the custom macros, operators are able to capture, index and store six documents per minute. At an average volume of 450 per day, the process requires just one to two hours. Images are cached temporarily and then committed to 5.25-inch WORM optical disks housed in a 24-disk jukebox, providing for a near-line image base of approximately one million pages.

Retrievals are made primarily by customer service people in the accounting department. Inquiries are divided about equally between customers and internal sources. Because of the ease of retrieval with the new system, substantially more inquiries are made now than with the manual filing system. The ease of retrievals is facilitated by the Advanced Optical Services (AOS) option of the MI^{3}MS 3000 system. For example, after images are retrieved they are temporarily cached in the event that other requestors may need them. The fax capability of the system is used to communicate images direct to requestors from the network nodes.

The system consists of:

MI^{3}MS 3000 version 2.5 multi-user software with five-user license
AOS (optical disk and jukebox management software)
MI^{3}MS 3000 I.C.E. (COLD software)

Novell network operating system with Ethernet backbone (a new network)
Minolta 750P file and database servers
DS 2400 document scanner
AC 1424 Autochanger with two internal 5.25-inch drives
Ten customer 486 PCS
Minolta 3100P Laser Printer

Benefits

An imaging system with advanced features such as compound document architecture, integrated computer output to laser disk (COLD) and AOS has resulted in a measurable increase in the Red Ball's productivity, saving the company approximately $30,000 in transaction costs to date. Filing is more accurate—the "file and misfile" label is history—and the files are more accessible to many more users, including people in the plant who would not have had access before.

The customer says that their biggest volume of paper is in this application, and they have made a big dent in reducing that volume. Aside from the achievement of strategic business goals such as reduced transaction costs, the reduced labor costs will result in a three-year payback.

Fast facts

Product Installed:	MI^{3}MS 3000 V. 2.5, 5-user license
Application:	Delivery receipt/invoice tracking
System configuration:	MI^{3}MS 3000 five-seat network; Advanced Optical Services (AOS) software; MI^{3}MS 3000 I.C.E. software; Minolta 750P file and database servers; Document Scanner, DS 2400; AC 1424 Autochanger with two 5.25-inch internal drives; Novell Netware with Ethernet backbone.
Operating System:	DOS/Windows
Network Software:	Novell Netware
Host integration:	None
Customization:	None
Other Products Used:	Advanced Optical Services (AOS) software
	MI^{3}MS 3000 I.C.E. software
Benefits:	Measurable increase in productivity—eliminated misfiles, permits multiple access to files, more accessible files. Reduced dependence on paper. Reduced transaction costs.

Trigon Blue Cross Blue Shield
Richmond, Virginia

North America Excellence Award: Imaging, Silver

1.) Describe the system application. What the system is used for, who are the users and what does the job entail?

The Users:

Trigon utilizes OPEN/image and OPEN/workflow technology for the automation of its Health Claims processing that supports its Strategic Business Units (SBU) and their approximate 65 associated Dedicated Service Units (DSU). The system supports more than 600 users (primarily Trigon's Health Claims Representatives (HCR) and Medical Advisors), with plans to extend to 1,000.

The Application: Health Claims Processing:

A typical health insurance claims is one to two pages in length and the system today provides access to more than 12 million documents electronically. The typical daily volume is 30,000 claims, or approximately 60,000 documents processed.

The Mail Room:

The process begins in the mailroom for the input of claims where they are scanned and OCR'd into the workflow and imaging system. Trigon researched the scanning requirements extensively during their feasibility study, and determined that like organizations were experiencing as high a rescan rate as 5 percent due to poor quality image display from the scanning process. Trigon researched and invested in "IPT" boards that provide a higher quality image during the scanning process, the result of which is a 99.7 percent image quality acceptance due to the additional investment. With additional mailroom training they've increased the through put of the OCR from 90,000 per month to more than 250,000.

Claims Processing:

Trigon, like other insurers, is required by law and regulatory requirements to pay interest on outstanding claims of more than 15 days, and are penalized for Medicare claims later than 31 days. The workflow push factor is a critical component for processing to meet their business goals. The rules-based workflows have facilitated new, more disciplined claims processing with embedded electronic control and tracking of work items in support of Trigon's business process.

Today, in the Proposition Three phase (detail provided in section six) of the system, once scanned and indexed, pending claims are triggered to workflow by the IBM mainframe and automatically distributed to the HCRs for processing. The claims are prioritized and forwarded to the correct DSU for processing and queued in the in-box by Julian date prompting the flow of claims to the HCR on First-In-First-Out (FIFO) basis, or by member priority.

Improved Productivity through PowerBuilder:

The claim is reviewed by the HCR and if necessary may be forwarded to a Medical Advisor for approval of treatment protocols and payment. Trigon has increased the productivity of the HCR in processing the claim with a unique graphical user interface developed with PowerBuilder. This user interface, coined "CINQ" for Customer Inquiry, automatically inputs data for the HCR via "screen scrapping" from as few as two, to as many as nine IBM mainframe based applications. CINQ has delivered dramatic keystroke reductions via automated system navigation and integration of code deciphering for routine transaction sequences. CINQ has reduced the number of keystrokes per claim for the HCR from a range of 50 to 300, to 25 or less in many circumstances. The automation of keystrokes, combined with the screen scrapping for the IBM based applications has also delivered an additional level of quality assurance in the claim process.

Multi-Window Claims Processing:

Prior to implementation of CINQ and the 21" monitors, the HCR would toggle back and forth on a 3270 screen to each of the two to nine applications as they processed the paper claim. The HCR would then make note of information and write it down many times on paper for input to the appropriate screen for processing. The CINQ application, and its screen scrapping functionality, has eliminated the need to toggle between applications and manually input data, CINQ now does this for the HCR automatically. The 21" monitors and multi-window capability allows them to view an entire claim on the screen for processing.

Enhanced Quality Assurance and 24-hour, Single Call Turn Around:

The system integration with the IBM mainframe application(s) also allows for the alert of any potential duplicate claim submissions— highlighted in red— based on coded data tracked by the mainframe(s). Prior to the workflow and imaging system the HCR would toggle back and forth in the screens to determine if the claim was suspect for redundancy, and then they would request a copy of the record(s) from microfilm and receive the information needed in 2-6 days. Today the system automatically routes and displays the images from each of the claims on the screen. The HCR is then able to assess if indeed the claim is a duplicate and previously paid.

Once the HCR has completed their analysis, workflow automatically routes to the appropriate Medical Advisor for approval or denial of payment, and explanation of benefits (EOB). Once reviewed by the Medical Advisor, workflow electronically routes the response back to the HCR, elevating its status in the in-box as priority work-in-process, for resolution with the customer. The system has eliminated the need to forward the information via interoffice mail systems with paper tracking cover sheets. The system has also eliminated the necessity for redundant filing systems by each of the HCR's as assurance against the loss of claim documentation as the paper copy was once routed throughout the organization.

As a result of the imaging and workflow system, many claims are now completed within 24 hours, and customer inquiries are addressed on the initial call eliminating the need for a call back. These processes could take 2-6 days depending on what additional information was needed, which would have been requested from the microfilm library via the IBM mainframe application.

Additionally, Trigon has implemented a database of the most-asked customer inquiries which has provided a higher quality in the consistency of how claims are managed, approved or denied, and how customer inquiries are answered. The database also provides Trigon the ability for workload balancing and to assist in expanding the role of the HCR to a multi-functional job design that allows them to support the processing of claims for other DSU's during peak periods. The HCR now has information available on screen for the processing of claims for a different DSU that assists Trigon in better managing its backlog when necessary.

Enhanced Job Structure & Employee Morale:

Of note is the value the system has brought to the HCR's and mailroom clerk's job functions. The HCR are now productive beginning at 7:00 a.m. in the morning. Prior to automation the Team Leaders would spend 2-2.5 hours in manually assembling and physically distributing the daily claims inventory for processing. The HCR's in the meantime would catch up on customer correspondence by handwriting letters and forwarding to the secretarial pool, and mailing those letters typed from the prior day's correspondence. Today their claims are automatically sorted, prioritized, and prompted based on Trigon's business rules through their in-box when they arrive. Their correspondence is automatically generated parallel with the claims processing through integration with word processing on their desktop. Additionally, bilingual HCRs can automatically translate the correspondence into Spanish providing additional competitive advantage in support of new business.

For the mailroom clerk, the system has provided the opportunity to develop expertise in computer scanning and indexing technology and a set of skills that provide for future job growth. The system has also allowed Trigon to shift the cost associated with Team Leaders assembling the daily claims inventory to the mail room clerk function. Also providing Trigon the opportunity to reclaim an additional two hours of their Team Leader's knowledge capability in processing claims and managing their teams, enhancing Trigon's profitability.

Competitive Advantage through Telecommuting:

Telecommuting has allowed Trigon to deliver additional levels of customer satisfaction with the ability to place Customer Service Representatives on-site at member organizations. Trigon has already implemented this "virtual office" capability through (Trigon on-line Paperless System) workstation at their 113,000 member program at the Commonwealth of Virginia, and the 13,000 member program for the Newport News Shipbuilding facility (see attached article). Trigon views telecommuting as providing them an additional level of competitive advantage.

Trigon's future vision for telecommuting includes the expansion to home-based claims processing, and the competitive ability to leverage their existing member sites when competing for new business opportunities. Other benefits of telecommuting that the Pilot Team identified included: the ability to expand on their available labor pool opportunities, lower labor costs, place satellite offices in less expensive rural locations, and the ability to alleviate some space constraints as they exist in Richmond today.

Awareness of the benefits of telecommuting reached an even higher level of understanding with the challenging weather conditions of this most recent 1995/1996 winter. With the future implementation of home-based telecommuting, their claims processing would not come to a standstill, while the backlog increased with each mail delivery during inclement weather when employees cannot travel to the physical office sites. Additionally, telecommuting could be a further benefit in workload balancing during peak periods. Telecommuting could provide them the means by which to keep on top of the 15-day interest requirement for unpaid claims, and the 31-day turnaround required by Medicare for reimbursement, which became a most serious backlog this past winter impacting revenue streams and causing significant corporate headaches and concerns.

2.) What were the key motivations behind installing this system?

Trigon Blue Cross Blue Shield is the largest health insurer in the state of Virginia, with more than $2 billion per year in revenues. Trigon provides coverage to approximately 1.8 million Virginians, and is also one of the largest employers with about 3,800 full time employees, of which approximately 1,100 are in medical claims processing and customer service in the Richmond facility.

To maintain their market position and to gain competitive advantage, Trigon decided to investigate new technologies to streamline processes and improve turn around for medical claims processing instead of continuing to invest in outdated, slow to retrieve, and expensive to maintain microfilm technology. Their CIO drove the effort after a site visit to Sigma Imaging Systems in New York.

Additionally, they had identified the long-term direction for corporate expansion in Virginia by increasing utilization of the company's managed care products, entering into new markets such as Medicaid and Medicare HMOs, and forming collaborative relationships with provider groups and acquiring other managed care companies. The company also intends to expand outside of Virginia through a combination of acquisitions and strategic alliances with managed care companies, traditional indemnity companies, other health care providers and other Blue Cross Blue Shield companies.

The original consideration was first based purely on the replacement of microfilm based archives which delayed claims processing anywhere from 2-6 days as well as customer inquiry follow-up. To determine how best to leverage what was originally just imaging technology, Trigon put in place a Pilot Team to assess how they might further leverage technology for strategic business goals. This team assessed three different levels of implementation termed "propositions" for consideration by their executive management team.

The team also took into consideration the unique requirements that the acquisition of managed care facilities may have on the automation of documents. There would be a relatively high added intensity of documents within these managed care environments and the system would need a great deal of flexibility to address this. New document types were likely to emerge from the episode-based and outcomes-based product and reimbursement strategies being discussed in the managed care environment.

In addition, the team assessed how telecommuting could provide an additional level of competitive advantage. They identified the advantages and customer satisfaction benefits associated with on-site customer service offices at member sites, home-based claims processing, ability to expand on labor pool opportunities, lower labor costs, place satellite offices in less expensive rural locations, and to alleviate space constraints in as they exist in Richmond today.

3.) Please describe the current system configuration (number and type of software, servers, scanners, printers, storage devices, etc.):

Clients:

- 600 Pentium and 486 based workstations with 21" monitors.
- Two Remote Workstation Sites Supporting Customer Service for 126,000 members via ISDN technology

Scanners & OCR:

- Four 923 Kodak Scanners, processing approximately 60,000 documents per day.
- Recognition OCR throughput up from 90,000 per month to 250,000 per month

Storage:

- Four HP Jukeboxes supporting over 12 million images

4.) Describe how the company has been impacted by this system. Be as specific as possible:

- Business process design improvements that have leveraged the power of teams, and developed a more streamlined, responsive organization with multi-functional job redesign that has provided Trigon more flexibility in the enhancement of customer satisfaction levels as well as employee job satisfaction and morale.
- On line Explanation of Benefits (EOB's) and database of most frequent inquiries has empowered workload balancing across the DSUs.
- Competitive advantage and expertise gained with on-site telecommuting Customer Service Representatives at large Member Service Organizations. Trigon has implemented this today at the Commonwealth of Virginia, a 113,000 member service site, and at the Newport News Shipbuilding, a 13,000 member service site.
- Opportunity to expand labor pools, and defray workload balancing.

4a.) What cost savings or increased revenues have been realized since the system was first installed?

- Thirty FTE (Full Time Employee) avoidance at $32,000 annual benefit rate totaled $960,000.
- A 24 percent internal rate of return was projected, a 34.3 percent internal rate of return has been realized.

- Microfilm cost avoidance: $279,450 camera replacements, $21,000 reader replacement, $50,000 annual equipment maintenance, $40,000 annual supplies and service.
- Expanded use of OCR— enabled service bureau cost avoidance from Service Bureau Keying: $240,000 annually

4b.) What productivity improvements have been realized?

- Productivity increased by a net 15 percent for those Health Claim Representatives with imaging and workflow on their desktops.
- Identification of a Power User/Team Leader within each implementation unit was key to successful reengineering and start up.
- The 21" monitors provide more concurrent and reasonably sized windows into the corporate systems running on IBM mainframes. The CINQ PowerBuilder application screen scrapes information for the HCR's eliminating 50 to 300 keystrokes.
- Triggering of the workflow process by the IBM CHIPS system ensures FIFO processing of claims and ensures protection against any related regulatory impact.
- Health Claims Representatives are now focused on the information delivered on the screen versus paper on the desk.
- Workflow-enabled exchange between the Health Claims Representative and the Medical Advisors has realized an average 24-hour turnaround for closure of claims vs. 4-6 days with manual process. HCRs can quickly annotate and forward documents while ensuring work-in-process integrity (no lost documents) and copying. The productive burden of counting, routing, batching, aging, sorting, searching, annotating, etc., by Team Leaders has been dramatically reduced.
- Customer inquiries are handled immediately on the phone by the Health Claims Representatives with access to information on Explanation of Benefits (EOB) provided by the membership plan and explanation of coverage from the Medical Advisors.

5.) Describe the implementation process and methodology, the project team, and any change in management and business process reengineering issues addressed:

Reengineer or automate?

Trigon chose to reengineer business processes prior to automation. The decision was based on the variety of needs for each of the Strategic Business Units (SBU), and their approximate 65 Dedicated Service Unit (DSU). Each operational division has different philosophies, practices and financial circumstances. Each of these SBUs required flexibility in how and when they might exploit the enabling PC, imaging and workflow technologies.

The pilot team also believed it to be in the best interests of the company to have a common document archive, versus potentially multiple system implementations by any of the SBUs or their associated DSUs. The team determined that it was critical to their environment to apply a redesign-first-then-automate mind set. The reengineering goals included the elimination of many hand-offs in the paper-based process, redesign for multi-functional job responsibilities, employee empowerment and increased job satisfaction, while capturing all information at the source.

With the needs for flexibility and a common document archive, and a redesign-first-then-automate mind set as their primary drivers the pilot team determined that a client server environment was the architecture of choice and then developed a three-proposition proposal for implementation as having the highest level of success for Trigon.

The Implementation Process and Methodology:

Proposition One: Implement an Imaging Pipeline:

The first proposition recommended a Richmond-based imaging pipeline to replace the antiquated microfilm equipment and avoid any additional investments in this technology. The pilot team realized that in order to encourage end users to utilize the system it was determined that an imaging pipeline would be the first phase in development. This phase, while rudimentary, was a critical-to-success factor to move individual endusers from the paper/microfilm environment to the PC/image display technology, a fair amount of claims needed to be accessible to jump-start the learning curve. The following were determined as critical factors to this phase -Proposition One of the implementation:

- the service of a "Power User/Team Leader" within each DSU implementation is a necessity.
- developed and implemented the infrastructure for digital images to replace microfilm as the
- archive of record for claims and related documents in Richmond
- four high speed image scanners placed into production in Richmond
- researched and employed automated image quality enhancement technology to dramatically
- reduce image re-scan issues. Currently experiencing image quality in excess of 99.7%
- implemented "forms drop out" technology for efficiency in optical storage and network
- utilization - forms reappear to end users.
- pipeline paradigm provides customers a broad spectrum of choices in how to exploit image
- capabilities for business results improvement. Customers may connect to the pipeline only for archive-only accesses, or may implement sophisticated workflows integrated with other systems.
- implemented Propositions One and Two in the first quarter of 1995 with a 17 percent internal rate of return in Richmond

Proposition Two—Redesign Document Capture Processes:

Proposition Two identified that there was an excellent opportunity to improve the quality and productivity of Trigon's document and data capture processes. The pilot team identified reengineering of these processes could significantly reduce hand-offs of paper within teams, reduce the number of keystrokes and ensure a reduction in mis-keystrokes in the health claims processing. The document capture would also ensure that claims would be processed by date-of-service routing and a strengthened commitment to OCR/ICR technology would be quite attractive operationally and financially. The team analyzed the combined savings of Propositions One and Two to be slightly positive on the financial entry into the document archive process and that it would have significant operation and service potential.

The implementation was also designed to be flexible. The pilot team's goal was to provide each business unit a means to implement so they could make informed and financially wise choices as they became acclimated with these enabling technologies. The flexibility of implementation provided options so that one workgroup could elect to provide relatively few image/retrieval/print workstations provided by the proposition pipeline completion. Another workgroup might elect to implement desktop delivery of images for only exception claims processing or only for archive access. Other workgroups might determine to fully exploit automated workflow to provide more robust and automated workflow management capabilities and aggressively eliminate paper.

The following were determined as critical factors to Proposition Two:
Reengineer front-end processes to include a multi-functional job design

- Incorporation of inventory building (claim) tasks, and a renewed commitment to exploit OCR/ICR technology.
- Work process redesign to reduce hand-offs and employ teams
- Implement Easel Inventory Enable (EIE) graphical software to deliver date-of-service specific document routing, and concurrent computer-assisted CHIPS (IBM) claims inventory building.
- Design and development of graphical applications that will deliver significant key stroke reductions, system navigation assistance, and code deciphering for routine transaction sequences.
- Eliminate CHIPS inventory building tasks by Team Leaders/Knowledge Workers in claims processing area.
- OCR usage level to expand from approximately 90,000 claims per month to 275,000 claims per month.
- Implemented "data staging" capability to provide for flexibility in how electronic data flows through the CHIPS system

Proposition Three—Leverage Trigon On-line Paperless Systems, "TOPS"

Implement image-based automated workflows within operations areas based on acceptance of business cases:

- Provide automated routing, measurement and aging through automated workflow prioritization

- A lead Business Unit was selected to lead the charge in the reengineering initiative, empowering all its team members with TOPS workstations and the resultant productivity and service enabling tools
- Proposition Three is providing the company with a 34 percent internal rate of return
- A Business Unit has also implemented remote TOPS services at Commonwealth of Virginia and Newport News Shipbuilders. Remote workstation innovatively provides a valuable learning curve for probable future remote graphical applications, satellite processing centers, or at home processing capabilities.
- A follow on Business Unit has completed installing a TOPS workstation
- Implement additional rules-based automated workflows by document type
- Continue implementation of Automated Mainframe Prompting to Workflow of Pending Claims
- Exploit facsimile capabilities for improved control and service
- Explore the business case for service scripts and improve productivity reporting
- Continue reengineering and reduction of pending cases

End-User Training

Trigon's approach to their reengineering efforts incorporated close involvement of their end-user community and their identified Power Users/Team Leaders from each of the DSUs. Prior to implementing a training program Trigon surveyed the end-users as to how best meet their needs. The response was "give it to us in manageable chunks, in a phased approach and build upon what we learn with each subsequent session." Trigon has since implemented a three phase training approach. They have also introduced a monthly Employee Newsletter highlighting productivity tips and most asked questions and answers, as well who to contact with additional questions and assistance.

Phase One Training:

As each DSU is scheduled to come on-line, Trigon provides them with a full day of PC Basics and Windows training. The goal is to assist them in transitioning from their 3270 dumb terminals to the 21" monitors with the ability to view multiple windows at once. Parallel to this first day of training, Trigon's Pilot Team will remove all of the endusers' 3270 workstations and replace with optimum-configured 21" monitors.

Phase Two Training:

As the DSUs become comfortable with the PC and Windows technology a second half-day training is scheduled where they are introduced to CINQ, the graphical user interface PowerBuilder application. CINQ automatically navigates across systems and screen scrapes the data directly into the claims processing effectively reducing what once took 50 to 300 keystrokes to accomplish to less than 25 in most cases. Trigon continues to develop value-add Macros with the Power User/Team Leader for further automation of keystrokes.

Phase Three Training:

The final phase of training where the PSUs are provided 1-2 hours training on the workflow in-box coupled with the integrated imaging.

The following table demonstrates how the "Proposition's 1-3" and the Three Phases of Training are implemented, and how it is rolled out to the Strategic Business Units and their associated Personalized Service Units:

SBU	Lead Business Unit	Follow on Business Unit	Remote Business Unit
Technology	Proposition Three	Proposition Two	Proposition One
Workflow Half-day Workflow In-box Training.	X		
Workstation Usage Half-day PowerBuilder CINQ Training. M/F Integration Screen Scrapping	X	X	
Archive Only One Day Windows Training. Removal of 3270 Workstation. PC Avail Next Day	X	X	X

Claims Process Engineering Progress Timeline

1992-1994
Extensive Workflow analysis within Shared Data Preparation and Lead Business Units with focus on reengineered process for improved quality, customer service, reduce lost documents and save time and money
Ongoing consultations with industry experts to explore the highest quality image capture, indexing, and distribution scenarios
Detailed financial models of both a pilot and large scale implementation, defining budgetary and cash flow expectations
Analyze and Identify Strategic Architectural Platform (UNIX, Novell, OS/2)

Feasibility Study, Technology Architecture Decisions, Pilot, Exec. Level Financial Buy-In

	1994	
Start of expanded OCR	January	OCR usage currently at 90,000 per month
	February	Document capture redesign approved
Management go-ahead for image pipeline	April	Team training completed for DMS depart.
	May	Management go-ahead for Lead

		Business Unit workstations
	June	Data & electrical rewiring of LBU dept.
Inventory Enabler program implemented	July	OCR usage exceeds 250,000 per month
	August	Completed document management. office renovation
Associate training completed & install of 225 workstations in LBU department	September	225 Workstations Installed!
LBU begins use of macros, WP, etc.	October	Implemented Richmond imaging pipeline
Start image archive with LBU dept.	November	LBU exploits "Customer Service Assistance" Program via PowerBuilder Front end
Executive Level go ahead for automated prompting	December	One million images on line!
	1995	
	January	Implemented ADR Imaging APPC Interface
Piloted ISDN/offsite connectivity	February	Implemented remote BU imaging pipeline
Executive Level go ahead for the Follow On BU TOPS implementations	March	Three million images on line!
		Trigon Financial Dept. establishes "Internal Rate of Return" analysis at 34.3 percent exceeding pilot estimate of 25 percent.
		Reduction of Keystrokes from 300 to 25 or less.
	April	Ad Hoc Workflow begins
Begin TOPS installation In Follow On BU.	May	4.7 million Images on line.
Remote TOPS Telecommuting installation at Newport News Shipyard for 13,000 Member Services pilot program	June	Live "TOPS" at Newport News Shipyard
	July	
Complete TOPS installation in Follow On BU	August	6.7 million Images on line.
		600 Workstations Installed.
		Start of Automated Workflow and Image enabled processes in Major

Remote TOPS Telecommuting installation at Commonwealth of Virginia for 113,000 Member Services Program	September	Retrievals sub 30 seconds, ISDN costs approx. two cents per minute
	1996	
	January	
Provider Network Management TOPS pilot begun	February	11.5 million Images on line.

Yarra Valley Water
Melbourne, Victoria, Australia

Asia Excellence Award: Workflow and Imaging, Gold

Executive Summary

Corporatizing the supply of water in Australia's second largest city and breaking away from an expensive mainframe environment provided the impetus for water supplier Yarra Valley Water to develop and implement image and data retrieval systems in a cost-effective client/server environment.

When the Victorian Government restructured Melbourne's water industry (Melbourne Water) on 1 January 1995, it was evident the three retail water companies formed by the split— Yarra Valley Water, City West Water and South East Water— would have to sever ties with the archaic and expensive mainframe-based system owned by the parent company. This system was costing Yarra Valley Water a massive $A10 million a year to support. In addition, any customer query took days to be answered.

Yarra Valley Water was determined to turn that situation around. This was an opportunity to develop an IT strategy that would fulfill its service delivery objectives of increasing customer service and profitability while reducing costs. It also provided them with the opportunity to excel against their competition in meeting the Victorian Government measurement criteria against which each of the three are to compete and be measured and ranked upon:

- Level of customer service and satisfaction
- Financial bottom line performance
- Environmental capabilities

Installation

Yarra Valley Water moved to a client/server-based environment using Sun servers and PCs. This new client/server platform reduced the annual support cost by more than 50 percent to an approximate $A4.9m per annum. The platform was used as the basis for the installation of Wang Laboratories, Inc. (Wang)'s OPEN/Image™ sofware and OPEN/coldplus™ sofware (computer output to laser disk) products which enabled further systems integration for an electronic workflow environment for internal business management and customer requests. In this environment, Yarra Valley Water developed "EasyAccess" and "Propertyflow" which include Wang's OPEN/image, Staffware's workflow product, to assist two of Yarra Valley Water's vital customer groups— plumbers and lawyers.

These customers use Yarra Valley Water's property sewerage plans and associated mapping information as part of their day-to-day plumbing and conveyancing services.

Not surprisingly, plumbers and lawyers want quick and convenient access to property information. The short turnaround of this information has allowed them to be more flexible in servicing their end customers.

Using EasyAccess, Melbourne plumbers now receive property sewerage plans and approvals from their local plumber supplier in less than 15 minutes. With Propertyflow, lawyers send requests by modem and have a conveyancing package delivered back within a time period as short as 30 minutes if requested. These times are a vast improvement on what customers previously faced— 10 days and two weeks respectively.

Each customer service system is integrated with the OPEN/image system which interfaces with Yarra Valley Water's workflow mapping, billing and document management systems for the quick collation of information requested by customers. Wang's OPEN/coldplusTM software is utilized for on-line financial and billing history.

The benefits to plumbers using EasyAccess include:

- Enhanced client satisfaction levels through the speed of process realized by their own end customers;
- The automation of the way plumbers do business with Yarra Valley Water by reducing the time spent processing standard applications from ten days to less than 15 minutes;
- Greater access geographically to services and property data via any one of ten plumbing suppliers who are linked to the Yarra Valley Water computer system;
- The relevant approval, copy of the property sewerage plan and any other information such as special conditions of connection are automatically sent back electronically to the plumbing outlet without any manual intervention from Yarra Valley Water;
- Paperwork is reduced and plumbers can pay on-the-spot for the service;
- Plumbers realize the added convenience of picking up supplies while waiting for the application to process;
- EasyAccess is available to plumbers outside regular business hours.
- Reduction in operating expense of an approximate $A3, 000 per annum as a result of the improved service provided.

Benefits to lawyers using Propertyflow include:

- Ability to generate a faster, more efficient method of providing property information to the conveyancing industry;
- Enhanced client satisfaction levels through the speed of process offered;
- Standard requests have been reduced from two weeks to a 30 minute response time if requested;
- The incorporation of a billing module for convenient monthly customer accounts;
- Money back guarantee for service delivery;
- Reduction in operating expense of an approximate $A3, 000 per annum as a result of the improved service provided.

In-house benefits to Yarra Valley Water include:

- An approximate 50 percent reduction in per annum system support costs (approximately $A4.9M), as compared to the prior $A10M per annum for mainframe support;
- EasyAccess and Propertyflow generating $A1.2M, a 40 percent increase in revenue per annum for the products;
- Work management solutions have realized Yarra Valley Water an 8.5 percent reduction in staff for the overall company, realizing $A1.3M salary saving per annum;
- Efficiency savings passed on to customers of $A3 per application, an approximate $A3,000 savings per customer per annum;
- Significant gains in customer satisfaction levels, which has also provided a "food chain" effect to their customer base whose own clients are also realizing higher levels;
- System ability to be transplanted to any number of government utilities such as water, electricity and roads, offering greater efficiencies to the public at large, and additional future revenue stream potential for Yarra Valley Water;
- The system is measurably allowing them to meet the ranking goals as stated by the Victorian Government: customer service and satisfaction, financial performance, environmental betterment.

In-house benefits to Yarra Valley Water using OPEN/coldplus include:

- Monthly financial reports have gone from being presented as bulky paper documents to real-time on-line data;
- Managers no longer wait for several days for reports, but have immediate online access to financial, billing history and ratepayer correspondence;
- Response time across 4.4 million pages of information is within 1.5 seconds;
- Management efficiencies include tighter controls of financials and internal resources;
- Access to seven years of data on-line.

Substantial Benefits: Access by plumbers and lawyers to vital property information has been substantially increased making Yarra Valley Water's service to its customers not only more efficient, but enabling substantial savings and workflow efficiencies for all concerned. Customers receive a greater level of customer satisfaction because of the easy to use front-end system. It is tightly integrated with all other systems, enabling immediate feedback and billing.

System Scope: Yarra Valley Water is responsible for supplying water to more than 600,000 customers in the Melbourne area. Any plumbers servicing the region can access approvals, property information and any other information electronically from EasyAccess outlets at plumbing supply stores. Any lawyer servicing the area can make inquiries to Yarra Valley Water's Propertyflow application— via modem, mail or fax. Internally, both Wang's OPEN/image and OPEN/coldplus software have been licensed to 100 users with appropriate access rights. Installed this year 1996, the system currently maintains 1.5 million customer-related documents. Additionally, the new electronic fee-based services are generating $A1.2M per annum, or a gain of 40 percent in revenue for the same products in the prior period.

Technology Innovation: Yarra Valley Water assigned Wang and Staffware the job of developing alternatives to nominated applications resident on the legacy mainframe hardware. Wang and Staffware integrated the output from several of Yarra Valley Water's existing core applications and developed packaging and delivery mechanisms for customer service applications. Wang provided the imaging system, OPEN/image, and computer output to laser disk system, OPEN/coldplus. OPEN/image system holds Yarra Valley Water's key imaging data such as property sewerage plans and agreements and indemnities for over 600,000 properties. This system resides within the Yarra Valley Water Company, but is accessible to plumbers and lawyers. In this first year of electronic access, plumbers alone will processe approximately 10,000 applications, or 50 percent of all transactions by plumbers, through EasyAccess.

The EasyAccess solution was developed using Wang's Open/image and Staffware's workflow software for plumbers to efficiently lodge plumbing applications which involved drawing on property information held by Yarra Valley Water. The Propertyflow system, also integrated by Wang and Staffware, enables Yarra Valley Water to respond quickly to lawyers' requests for property sewerage plans, encumbrances, rate assessments and images of any indemnities or agreements on the property.

Wang also integrated its OPEN/image and OPEN/coldplus systems to Yarra Valley Water's financial system. This gives Yarra Valley Water's business managers immediate on-line access to financial and billing information. Yarra Valley Water can, within seconds, go back seven years to view customer-billing records. Financial information can be shared, modified and checked before monthly and end-of-year financial reports are electronically produced. This saves time and paper.

Implementation Approach: Wang Australia and Staffware were called in to discuss options and tender solutions to assist in developing a strategy that would enable Yarra Valley Water to move to a more cost-effective computer platform with improved customer service.

Part of this strategy involved developing innovative solutions to make disparate pre-existing applications work together. The implementation team consisted of Yarra Valley Water staff, plumbing and lawyer customers, Wang, Staffware and other system solution providers.

Yarra Valley Water provided very broad specifications in the first place. This was important to enable customers to have a high level of input into the final solution. It was realized that without customer input the systems could potentially have a disappointing level of functionality and not meet customer needs.

A roundtable approach with all interested parties meant that the original specifications were continually modified and added to as its contributors realized what was achievable. Interestingly, the client and the customer's expectation of what could be achieved in the final solution greatly increased. The more they discussed what they needed, the closer they came to achieving a very useful customer support system which was tightly integrated with Yarra Valley Water's business requirements.

Once the team established database requirements and data streams, it then undertook a prototype program. This was a pilot situation involving all users to test the initial validity of the system. Immediate user feedback was attained, and modifications made to provide early delivery of those features. Upon stabilization of the product Yarra Valley Water commenced marketing it to its customers in early 1996.

1.) Describe the system application. What the system is used for, who are the users and what does the job entail? How often or how many hours is the system in use on a daily basis.

Overview

To boost customer service, Yarra Valley Water utilized Wang's OPEN/image and OPEN/coldplus (computer output to laser disk) products.

These systems interface with Yarra Valley Water workflow mapping, billing and document management systems to collate the information requested by its plumbing and lawyer customers. The system providing direct property information to plumbers is called EasyAccess and to lawyers, Propertyflow.

EasyAccess

EasyAccess has automated the way plumbers do business with Yarra Valley Water by reducing the time spent processing applications. For example, to build an extension to a house that involves altering the plumbing, a plumber is required to lodge an application and obtain approval before commencing work.

With the old manual processes and legacy systems, it could take two weeks to process an application. Not only that, the plumbers had to come into Yarra Valley Water offices in person to sign the necessary forms and pay application fees up front. Once lodged, applications were checked manually, which took up a lot of resources internally. That would then be followed by phone calls inquiring about progress.

Now a plumber can go into any one of ten plumbing suppliers located over a wide geographical area to access the Yarra Valley Water computer system. After answering a few straightforward questions, the completed request is transmitted to Yarra Valley Water where a workflow case is automatically initiated.

The relevant approval, copy of the property sewerage plan and any other information such as special conditions of connection are automatically sent back electronically to the plumbing outlet without any manual intervention from Yarra Valley Water.

The service now provides a 15-minute turnaround for a standard request compared to ten days previously. Yarra Valley Water has simplified the process and then automated it so there is no need for manual checks of map bases, property, customer or plumber records. There is a lot less paperwork and the plumbers pay the plumbing shop a fee for the service. The result is that plumbers are keen to work with Yarra Valley Water and the level of service has improved dramatically. EasyAccess is also available to plumbers outside of the company's regular business hours.

Propertyflow

Yarra Valley Water has applied electronic technology to generate a faster, more efficient method of providing important property information to the conveyancing industry. Previously, a lawyer had to mail an application or send a clerk to Yarra Valley Water offices to make the application and pay for the service in advance. Handling requests manually by several different departments could take weeks before the package became available.

Now, inquiries come either via modem, mail or fax. Driven by Staffware's workflow software each item requested is accessed automatically from the Wang imaging system, for billing and encumbrance and mapping systems, and compiled in a file which can be either faxed directly or printed out and mailed to the lawyer. A standard request is for an eight-hour turnaround in a service that used to take up to two weeks. If you are buying a property at an auction, for example, and require a faster turnaround, the lawyer can request the information within half an hour for an appropriate fee. The service is also supported by a money-back guarantee if the information is not delivered in the time required.

The property information package includes evidence of any encumbrances, a rates assessment, including any arrears which will be part of settlement, a copy of the property sewerage plan, including sewerage pipes and drains, and an image of any indemnities or agreements on the property. The Propertyflow system also incorporates a billing module. At month's end, an itemized account of all services provided is sent to the customer.

Storage savings

Wang's systems integration has also enabled management efficiencies such as tighter controls of financials and internal resources within Yarra Valley Water's own operations. The OPEN/coldplus software, for example, contains records of current bills and billing history more than seven years old. In effect, the system compresses a print file to take out white space and standard forms, and stores the data.

Historically it has been very expensive to have all the necessary data on-line. It not only takes time to retrieve the document, but it can affect the on-line performance of core applications. OPEN/coldplus software is a much more cost-effective and efficient storage system.

When a user needs to access the financial history of a property or a customer has a query in relation to their bill, customer service staff can access the OPEN/coldplus system to retrieve the billing history. In the future, cold technology will be used to store electronic copies of bills so that both the customer and customer service staff can view the same document.

At each month's end in the mainframe days, managers would receive financial reports from the general ledger presented as thick wads of paper. Now management with appropriate access receives an electronic message telling them that the relevant reports are on-line. Rather than waiting for several days, they can look at the data reported at the end of the financial period immediately on-screen and compare, for example, 13 months of rolling financials.

Image Capture and Indexing: The initial image capture residing on the mainframe held around 700,000 images. These showed details about sewerage and water lines on properties. It was indexed by three fields. Wang wrote an application that enabled the image and index to be imported to the OPEN/image system for automated application processing. New properties in the Yarra Valley Water district are expected to grow at a rate of three percent per annum, with each new image of any additional property being captured and stored on OPEN/image software.

Pre-Process Coding: The nature of customer requests is determined when requests enter the OPEN/image system. Workflow software is used to automatically read and allocate requests. There are a variety of pre-determined requests that customers can make.

The workflow application designates which system within OPEN/image receives the request and automatically alerts the billing system to calculate the cost of the request.

Workflow Routing: The workflow software invokes Wang developed software to retrieve and package such information as property sewage plans, property maps, encumbrances, rate certificates and then presents them back to the user via modem, or in preparation for faxing or mailing.

Suspense Processing: There is no need to suspend processing as errors in customer request forms are detected and corrected on line.

Applicant Notification: Once the customer request is retrieved and automatically verified by Yarra Valley Water, it is then returned to the plumber or lawyer via fax, modem or mail.

Customer Service: Hours of operation are 7am-7pm weekdays and 7am-5pm weekends, for both Propertyflow and EasyAccess systems. If complications arise that require the intervention of a Yarra Valley Water officer, this service is available. However, such requests are virtually non-existent. The hours of operation will shortly be extended to continuous availability from 7.00am Friday to 6.00pm Sunday.

2.) What were the key motivations behind installing the system?

This was an opportunity for Yarra Valley Water to develop an IT strategy that would fulfill its service delivery objectives of increasing customer satisfaction and profitability while reducing costs.

When the Victorian Government restructured Melbourne's water industry into three retail water companies, it asked each retailer to compete and be ranked with each other on several criteria including:

- Level of customer service and satisfaction
- Financial and bottom line performance
- Environmental capabilities

Although Yarra Valley Water has not yet been rated independently on these criteria since it installed the new systems, strong anecdotal evidence indicates it has increased customer satisfaction over and above its competitors. The financial performance of the company has improved dramatically through enhanced work productivity and staff reductions. In addition, paper usage and customer time taken to travel to Yarra Valley Water has been drastically reduced, with positive impact on the environmental betterment agenda.

Yarra Valley Water's own surveys indicate customer satisfaction rose from 74 percent to 77 percent for all customers, including plumbers, lawyers and the general public. In addition, Yarra Valley Water has received many notifications thanking it for improving customer service.

Customers using the systems are receiving significant savings with reduced time in application lodging and travel times. Overall savings to each plumber and lawyer using the system is estimated to be approximately $A3,000 per year.

Yarra Valley Water unquestionably has become a primary water utility that, if privatized, would provide an attractive venture.

Further, Yarra Valley Water has developed a system which has the ability to be transplanted to any number of government utilities such as water, electricity and roads. As such, Yarra Valley Water and Wang are in discussion with other government utilities that have shown a keen interest to deliver similar cost-efficiencies and improved customer services.

3.) Please describe the current system configuration (number and type of software, servers, scanners, printers, storage devices, etc.).

Sites:	One— Yarra Valley Water
Image Systems:	One Sun Solaris server running Wang OPEN/image and OPEN/Cold^{plus}, Microsoft Windows clients.
	A further 11 servers running billing system, account system, asset management system and office automation system.
Volumes:	Plumbers: EasyAccess— 12,500 transactions per annum, with expected 100 percent growth rate over 6 months.
	Lawyers— Propertyflow— 40,000 per annum.
Operations:	7 am-7 pm weekdays, 7am-5pm weekends.
Scanners:	One Bell Howell 338 scanner.
	One Fujitsu 3096 scanner.
Workstations:	EasyAccess 10 PCs at plumbing outlets; 8 Yarra Valley Water PCs. Propertyflow 50 PCs at lawyer offices, with expected growth rate to 100 PCs in next six months; 5 Yarra Valley Water PCs.
Fax Gateways:	Dual Zetafax servers supporting 8 lines each.
Image Server:	One Sun 1000E4 server with 512Mb memory, 132Gb of hard disk, running Solaris and Wang OPEN/image and OPEN/^{coldplus} software.
Printers:	Five non-dedicated printers.
LAN:	10Mb Ethernet LAN.
Communications:	Via Ethernet and client/server.

4.) How is this system integrated with the company's other information processing systems?

There has been an enormous amount of integration involved in the provision of the Yarra Valley Water customer service and internal financial solution. Yarra Valley Water has integrated all database applications in order that workflow is automated and that information is passed electronically to appropriate areas of the company for business efficiencies.

Plumbing and lawyer inquiries on a property trigger the OPEN/image system to retrieve information. Databases that have been integrated with the OPEN/image system include the billing system, graphical information system and the encumbrance system.

Once a customer information package has been collated, the Wang developed software creates an entry to be posted to the core billing system to invoice the customer. The package will be passed to the document management system for archival so that history on that property can be retrieved.

At the end of the fiscal year, Yarra Valley Water produces reports on each ratepaying property. That information can be electronically transferred from the core billing system to the OPEN/^{coldplus} system that generates ratepayer bills for water, sewerage and refuse.

5.) Describe how the company has been impacted by this system. Be as specific as possible. What cost savings, increased revenues or work productivity improvements have been realized since the system was first installed?

Yarra Valley Water measures system benefits in three ways:

1. Improved customer service and satisfaction.

Since Propertyflow became operational to lawyers, approximately 50 high-volume lawyers, representing around 15 percent of lawyer volume, have begun using the system by modem or have applied to be linked by modem.

Approximately 50 percent of all lawyers in the area are either using Propertyflow via modem or facsimile.

Approximately 70 percent of applications by plumbers can be done using EasyAccess. Since EasyAccess became operational to plumbers, around 40 percent of all transactions that can be conducted by EasyAccess are being done by this system. These applications can be processed automatically without any Yarra Valley Water staff involvement. Additional services are planned that will increase the transaction rate at plumbing shops.

Response times to customers have vastly improved. Applications to lawyers previously took approximately two weeks. Applications now take between 30 minutes and three days, depending on the nature of the application and cost.

Application processing for plumbers previously took approximately ten days. Applications now take 15 minutes at a location that suits them. Plumbers no longer need to drive to Yarra Valley Water's main office to lodge an application.

Yarra Valley Water surveys indicate customer satisfaction rose from 74 percent to 77 percent for all customers, including plumbers lawyers and the general public. In addition, Yarra Valley Water has received many notifications thanking it for improving customer service.

2. Reductions in Yarra Valley Water costs.

There is estimated to be a dramatic reduction in system costs to Yarra Valley Water. The company is expecting to receive a significant return on its systems investment within 18 months. The previous mainframe system failed to make a return.

Reduction in Yarra Valley Water staff numbers has been 50 percent for plumber application processing and lawyer information processing. The total number is approximately 30 people. Yarra Valley Water has realized an overall 8.5 percent reduction in staff for the company, representing $A1.3M in salary savings per annum.

3. Pricing restructures

Yarra Valley Water has adjusted its pricing policy for application processing which has resulted in an overall saving to the customer of approximately $3 per application, while EasyAccess and Propertyflow are generating $A1.2M, a 40 percent increase in revenue per annum for the products.

Customers are receiving significant savings with reduced time in application lodging and travel times. Overall savings to each plumber and lawyer using the system is estimated to be approximately $A3,000 per year.

How has business workflow been affected (compared to before system implementation)?

Previously plumbers and lawyers had to lodge their inquiry, including payment, by mail or in person to Yarra Valley Water's main office. Yarra Valley Water manually processed each application using data entry staff. This could take up to two weeks.

Now lawyers dispatch applications from their office either via modem, mail or fax. Electronically lodged applications enter Yarra Valley Water's Propertyflow system and are automatically processed and billed. Information processing has been integrated and streamlined for those applications lodged by mail or fax. Once in the system, the application is automatically processed, then quickly checked, before being returned to the lawyer by fax or mail. This process now takes 30 minutes to three days (reduced from two weeks), depending on the nature of the application and costs.

Yarra Valley Water established a franchise among ten plumbing supply outlets for plumber applications. Plumbing outlets were chosen to house the system as a convenient location to plumbers wanting supplies.

Today the plumbing retailers undertake the data entry. The application is a simple checkbox application that takes around two minutes to enter and dispatch to Yarra Valley Water. The approval is given and faxed back within 15 minutes (reduced from 10 days). The OPEN/image system is automatically activated to release requested property sewerage images. Applications are processed without any Yarra Valley Water staff involvement. The plumbing retailer bills the plumber and Yarra Valley Water bills the retailer.

Yarra Valley Water has technical operators that can assist in more complicated plumbing jobs. It also has sales representatives who visit plumbing retailers to ensure the system is running smoothly. Yarra Valley Water staff morale is higher with greater customer involvement other than performing data entry work only.

6.) Describe the implementation process and methodology, the Project Team, and any change management and business process re-engineering addressed.

Methodology

Step 1— Yarra Valley Water and Wang established the database requirements. It was felt that the data elements should be independent of the business process to allow new ways of organizing data streams.

Step 2— Rapid prototyping was undertaken as proposed revised changes to business processes were tested.

Step 3— The Yarra Valley Water and the Wang and Staffware teams worked directly with key Yarra Valley Water customers to attain immediate feedback on those prototypes and to provide early delivery of those features. This was a pilot situation to test the initial validity of the system.

Step 4— The team documented the requirements on the basis of the prototypes.

Step 5— A system test cycle before formal hand over of the software to Yarra Valley Water.

Project Team

There were seven Wang, Staffware and third party specialists in the project team. This was a very fluid arrangement with many part-time participants and substantial customer input. In addition, there was continual informal interchange with participating vendors. Wang had the task of integrating the solution and delivering imaging and customer service products. Yarra Valley Water coordinated the project and delivery was reliant on the goodwill of each participating party.

Change Management and business process re-engineering

Change management procedures were an agreed process based on Wang's quality management system. This allowed users to ask for modifications to the system and nominate changes. Yarra Valley Water would then sign off prior to any changes being made. Wang used software that tracks its change management procedure. The software also tracks problems and records resolutions.

Business process reengineering was inherent in delivering the solution. The project team knew it had to think outside the existing Yarra Valley Water businesses process boundaries to improve the flow of information.

1.) Innovation

1A: Innovative use of technology to further strategic objectives

Propertyflow: The emphasis has been to push work from Yarra Valley Water out to lawyers' offices. This way Yarra Valley Water could reduce staff numbers and give lawyers more control over their property applications. Previously lawyers would fill out a form, send it off to Yarra Valley Water with a check, and wait for a reply. Lawyers wanted more control because they are often under pressure with property settlement issues. Now that they have their contribution to a conveyancing package automated, the next step is to offer similar services to other utilities in the property industry.

EasyAccess: Yarra Valley Water wanted to reduce the need for plumbers to visit Yarra Valley Water office for plumbing approvals and property plans. Now plumbers can have their applications processed while picking out supplies from plumbing shops. The plumbing retailer has been trained by Yarra Valley Water to handle applications and accepts fees on behalf of Yarra Valley Water for the service. EasyAccess handles standard plumbing cases only (approximately 70% of all plumbing transactions) at the moment. Wang and Staffware are involved in the development of non-standard cases that are currently being tested.

1B: Degree of complexity in the underlying business process and IT architecture. Complexity in the underlying business process involved:

In automating Yarra Valley Water's work processes, many mental barriers needed to be broken down. Many practices at Yarra Valley Water were extremely old. For example, lawyers had to pay their account before they received information and plumbers had to sign all application forms. The end result is that organizational boundaries have been broken down electronically. It used to be that several different departments would be involved in putting information together for customer information packages on topics such as property sewerage plans, drainage and encumbrances. Now this is completely automated or handled by a single person.

The EasyAccess & Propertyflow solutions have been in operation since early 1996 and Wang has been asked to provide further solutions in addition to the original specifications. In particular Wang has become involved in the preparation and customization of standard business letters as customers contact Yarra Valley Water by mail or phone.

Complexity within the IT architecture included:

Previously none of Yarra Valley Water's core business systems, including mapping, accounting and asset management systems, had ever worked together. The complexity arose out of integrating these pre-existing systems with each other and also integrating new systems, including Wang's OPEN/image and OPEN/coldplus software technology.

A further complexity entailed the need to communicate with all the vendors who had provided new and pre-existing hardware applications systems. In integrating the solution and developing integrated customer service systems, Wang needed to utilize technical expertise and effective communication skills with users and vendors to make the right decisions, while maintaining high morale among each participating party.

The following applications were integrated:

- Staffware workflow software and customer interface solution;
- the billing system with read-only customer details, rates and outstanding notices;
- the Phoenix system which contains maps of Yarra Valley Water assets such as water pipes on properties;
- Wang OPEN/image with property sewerage plans showing water and drainage pipes as well as any agreements and indemnities;
- Wang's Customer Information Exchange database featuring encumbrance information and Propertyflow and EasyAccess transactional information;
- Wang OPEN/coldplus running financial, billing history, ratepayer correspondence and document management systems.
- Some databases were non-compatible, which involved acquiring and installing database gateways for communication between disparate systems.

1C: Creative and successful deployment of advanced workflow and imaging concepts

- Activities/Use showing creative emphasis on imaging:
- Wang's OPEN/image allows Yarra Valley Water to store a variety of image types as part of EasyAccess. Propertyflow images from the OPEN/image repository are packaged with the output from otherwise linked systems for convenient delivery to Yarra Valley Water's customers.
- Activities demonstrating tangible success of imaging:
- If Yarra Valley Water had not undertaken the imaging of its property sewerage plans on every property in the metropolitan area, it would not be able to offer the Propertyflow and EasyAccess to its customers.
- Both the Propertyflow and the EasyAccess system operate from the property plans or the indemnities and agreements for properties. Customers need this information to be able to conduct their work. It was critical that the plans were scanned and imaged on Wang's OPEN/image system for EasyAccess to customers.

1D: Achievement of Business Process Reengineering and/or continuous improvement

Business process reengineering achievements were to reduce Yarra Valley Water staff levels and shorten the time frame in providing information to customers. The advantage in having workflow software at the front-end of the system is that this provides statistics for measuring performance of the system over time. Yarra Valley Water has developed these systems keeping in mind key public indicators of customer service and environmental betterment.

2.) Implementation

2A: Successful implementation approach

The emphasis was on prototyping user needs, formalizing the requirements as they evolved, testing the system, and then packaging and releasing the software to Yarra Valley Water. Inherent in this approach, which ultimately resulted in a successful solution, was discourse with users and vendors from the very beginning of the project.

2B: Extent of change management process

User Education:

Lawyers— Propertyflow system: All lawyers on the Yarra Valley Water database received an information package about the installation of the Propertyflow system. Those lawyers that indicated they would link up to the system by modem were offered training by Yarra Valley Water staff in their offices. This training took approximately 30 minutes.

Plumbers— EasyAccess: All plumbers on the Yarra Valley Water database were similarly mailed an information kit notifying them of the EasyAccess system. There is an ongoing marketing process involving focus group events at the plumbing retail outlets, and across-the-counter information when plumbers come to the Yarra Valley Water site.

Yarra Valley Water surveyed all plumbers that came into the site to find out why they were not using the system. The results of this survey are being used to tailor the educational program to plumbers.

In addition, training of franchisees involved two days for at least two people, including a store manager, at the Yarra Valley Water site. Yarra Valley Water is continuing to provide on-going refresher training to the retail outlets on a needs basis.

Internal Yarra Valley Water staff training— Propertyflow: Before it was released for general use, Yarra Valley Water asked training staff to produce every application using the new system three times. This provided an opportunity for staff to learn and find faults with the system so that any glitches or inefficiencies could be overcome before it was operational. The final trial showed no failings that indicated a stable system.

Internal staff training— EasyAccess: There was an initial training period of one day for EasyAccess staff, which is continuing as new products, including increased index and customer retrieval functionality, are being added to the system. Refresher courses are available on as-needed basis.

System Documentation and Training:

Wang produced four system documents for the Yarra Valley Water system. The first three documents are mandated by Wang's software development methodologies, the fourth was requested by Yarra Valley Water.

- The Product Functional Specification for Wang's development methodology, included formal documentation of system requirements. This document evolved throughout the prototyping period.
- The Software Design Specification containing the internal system design.
- The External Interface Specification which stresses the interface points of the system. Wang uses the last two documents for software maintenance purposes and for its in-house training program.
- The Hand Over Document to Yarra Valley Water containing the site's operational characteristics for on-line operational support of the system. Yarra Valley Water help desk staff use this document for training.

2C: Level of overall system complexity

System Scope:	100 OPEN/image licensed users
	100 OPEN/Coldplus licensed users
System Processing Capacity:	One Bell Howell 338 scanner
	One Fujitsu 3096 scanner
	One Sun Solaris server.
	Five Hewlett-Packard LaserJet 4 printers.
	62,000 images for lawyer and plumbing transactions.
	600,000 site images, with growth rate of around 3 per cent per year.
System Distribution:	One— Yarra Water Valley.
	60 remote plumbing and lawyer organizations.
	Mix of dial-up and high-speed leased lines.
System Integration:	One Fujitsu mainframe system (now redundant.)
	Propertyflow and EasyAccess systems.
	One Wang OPEN/coldplus system— financials, billing history, rates records.
	One OPEN/image system— property sewerage plans, Propertyflow and EasyAccess processing, integrated to workflow software, billing and maps.
	Sun server platform, IBM compatible PCs for client platform.
System upgradeability:	Adequate capacity to handle estimated business growth. If applications and users outgrow existing systems, the client/server platform can be cost-effectively enhanced with new servers and PCs.
	Jukeboxes can be added to handle excess cold documents and images.

2D: Level of integration with other advanced technologies

There was major integration with a number of other vendors to make Yarra Valley Water's internal and customer service systems seamless to the end users.

The vendor technologies other than Wang's OPEN/image and OPEN/Cold^{plus} technologies include:

Staffware	workflow software and for the front-end solution;
Oracle	database supplier;
Sun	Solaris hardware platform;
IBM	PC clients;
EDS	core billing and collection application;
TSW	asset management system;
Computron	financial system;
ConsulCAD	graphical information system.

3.) Impact

3A: Extent and impact of demonstrated productivity improvements

Since Propertyflow became operational to lawyers, approximately 50 high volume lawyers, representing around 15 percent of lawyer volume, have begun using the system by modem or have applied to be linked by modem.

Approximately 50 percent of all lawyers in the area are either using Propertyflow via modem or email. Approximately 70 percent of applications by plumbers can be done using EasyAccess. Since EasyAccess became operational to plumbers, around 40 percent of all transactions that can be conducted by EasyAccess are being done by this system. These applications can be processed automatically without any Yarra Valley Water staff involvement.

Response times to customers have vastly improved. Applications to lawyers previously took approximately two weeks. Applications now take between 30 minutes and three days, depending on the nature of the application and cost.

Application processing for plumbers previously took approximately ten days. Applications now take less than 15 minutes at a location that suits them. Plumbers no longer need to drive to Yarra Valley Water to lodge an application.

3B: Significance of cost savings

There is estimated to be a dramatic reduction in system costs to Yarra Valley Water. The company is expecting to receive a significant return on its systems investment within 18 months. The previous mainframe system failed to make a return.

Reduction in Yarra Valley Water staff numbers has been 50 percent for plumber application processing and 50 percent for lawyer information. The total staff reductions were about 30 people.

Customers are receiving significant savings with reduced time in application lodging and travel times. Overall savings to each plumber and lawyer using the system is estimated to be $A3000 per year.

3C: Level of increased revenues, product enhancements, and customer service or quality improvements.

Improved Revenue/Product Opportunities:

Yarra Valley Water has commenced work to expand its system to automate non-standard plumber applications and has asked Wang and Staffware to increase functionality of customer systems.

Improved revenue opportunities for Yarra Valley Water include the opportunity to on-sell the intellectual property rights of their systems. Yarra Valley Water is currently in discussion with other utility providers about this.

Improved Customer Service:

For lawyers: Applications by lawyers previously took approximately two weeks to process. They now take between 30 minutes or three days, depending on the nature of the application, cost, and is supported by a money-back guarantee.

Now lawyers dispatch applications from their office either via modem, mail or fax. The application is automatically processed, then quickly checked, before being returned to the lawyer by fax or mail. Payments can now be made in a customary business manner after the service has been delivered.

For plumbers: Application processing for plumbers previously took approximately 10 days. Applications now take less than 15 minutes at a location that suits them. Plumbers no longer need to drive to Yarra Valley Water to lodge an application. From the one location they can access the company's services and purchase plumbing supplies.

Improved Application Processing Quality:

Previously Yarra Valley Water manually processed each application using data entry staff. Now most applications can be processed without staff involvement. If there is staff intervention, it is by one person alone, rather than numerous people handling different, specialized areas of the application.

The workflow front-end provides consistencies in data gathering, data exit and validation process and an audit trail to track the progress of every application.

Monthly financial reports have gone from being presented as bulky paper documents to real-time on-line presentations. Managers no longer wait for several days for reports, but have immediate on-line access to financial, billing history and ratepayer correspondence. Response time across 4.4 million pages of information is within 1.5 seconds.

3D: Proven strategic importance to the organization's mission

Yarra Valley Water recognizes that water is an essential commodity to society. As such, it is committed to providing world class water and sewerage services. Its EasyAccess and Propertyflow systems have enabled the organization to take giant leaps in delivering faster customer service in the provision of these products.

These systems have proven a difference can be made in the importance of improving customer services and have given it a base in continuing to provide quality solutions cost-effectively.

By operating in a corporate environment, Yarra Valley Water has been able to change its focus to customer service and core business savings.

3E: Degree to which the system enabled a cultural change within the organization

Yarra Valley Water culture has changed to provide a more significant customer focus. The company has demonstrated that it can provide top quality customer service using clever technology. Yarra Valley Water changed its thinking from believing that it had to have a hands-on approach to customer service. Those notions have given way to the knowledge that technology can offer customers what they want and even go beyond their expectations in providing excellent service. Customers now access to information at the locations that suit them achieving results in shorter time frames.

Yarra Valley Water staff now have new and more challenging roles as they are no longer involved in laborious paper processing and data entry roles, but are either liaising with customers, working with computer soft copy or negotiating for the use of the system with other utilities organizations.

By decommissioning mainframe data processing, Yarra Valley Water now has a cost-effective client/server environment enabling it to be customer focused. Its attitude now is to get the job done and to not let obstacles stand in the way.

3F: Impact of system on competitive positioning in marketplace

Melbourne water supply reforms introduced competition in the provision of services rather than customers numbers. An independent Regulator General is in the process of benchmarking the three retail suppliers. These benchmark ratings are due to be released in December 1996.

In the meantime, Yarra Valley Water believes it is leading the field in providing innovative services to customers. Feedback from lawyer and plumbing customers, who are free to operate all over Melbourne, indicate they want other water retailers to adopt the same services and have asked Yarra Valley Water to organize that its competitors adopt the same systems.

The following matrix highlights some of the key benefits:

Corporate Overview	• Yarra Valley Water • One of 3 retail water providers in Melbourne as a result of the Victorian Government restructuring • 1.5 million customers with approximately 560,000 properties connected to the water or sewerage system, & associated 600,000 site plans on-line (property, sewerage, & service access maps) • Opportunity for Yarra Valley Water to develop an IT strategy that would fulfill its service delivery objectives of increasing customer satisfaction and profitability while reducing costs. • Wang and Staffware designated for mainframe transition to c/s • Wang and Staffware developed EasyAccess & Propertyflow Apps. • Accessible from lawyers offices and geographically dispersed plumbing supply retailers for application processing • Significant increase in customer satisfaction levels with "food chain" effect to their own customers clients due to faster processing • System ability to be transplanted to other gov-

	ernment utilities such as water, electricity and roads potential future revenue stream • Customers requesting the other 2 water retailers automate to Yarra Valley Water system • The system is measurably allowing Yarra Valley Water to compete against the other 2 retailers based on the Victorian Government's ranking criteria— customer service and satisfaction, financial performance, environmental betterment
System Impact	• Realized approx. 50 percent per annum system support cost savings through client server architecture now $A4.9M per annum as opposed to prior (inherited) mainframe $A10.0M per annum • EasyAccess and Propertyflow generating $A1.2M, a 40 percent increase in product revenues • 8.5 percent staff reduction for company as a whole, realizing $A1.3M salary savings per annum • Efficiency savings of $A3 per application passed to customers representing an approximate $A3,000 per annum per customer • System ability to be transplanted to other government utilities such as water, electricity and roads potential future revenue stream • The system is measurably allowing Yarra Valley Water to compete against the other 2 retailers based on the Victorian Government's ranking criteria • Only one of the three water retailers to automate
Innovative use of technology to further strategic objectives	• Delivered application processing capabilities to the Lawyers office, shortening application processing time from 2 weeks to 30 minutes when requested • Delivered application processing for plumbers directly to geographically dispersed retail plumbing suppliers, plumbers can have application processed while picking out supplies, shortened application time from 10 days to less than 15 minutes • System available to plumbers beyond normal business hours now 7:00 am— 7:00 pm weekdays and 7:00 am— 5:00 pm weekends (soon to provide continuous weekend coverage) • Enhanced customer satisfaction with "food chain" to their customers' clients through faster delivery of service to their clients for improved

	Yarra Valley Water processing time • Operating cost savings for both plumbers and lawyers of approximate $A3K per annum • Enhanced employee job satisfaction, developed a diverse/flexible workforce with an 8.5 percent reduction of staff or $A1.3M in salary savings per annum • Product revenues up 40 percent to $A1.2M per annum • Approximate $A5M per annum system support savings from prior mainframe system • 1.5 million documents on line • 7 years of financials managed online through OPEN/coldplus
Wang work management software	• OPEN/image & OPEN/coldplus for Sun Solaris • 100 User licenses
Propertyflow Application/Lawyers	• Developed for needs of Lawyers • Ability to generate a faster, more efficient method of providing property information to the conveyancing industry; • Enhanced client satisfaction levels through the speed of process offered; • Standard requests have been reduced from two weeks to a 30 minute response time if requested; • The incorporation of a billing module for convenient monthly customer accounts; • Money back guarantee for service deliver; • Reduction in operating expense of an approximate $A3,000 per annum as a result of the reduced fee structure provided.
EasyAccess Application/Plumbers	• Developed with input from Plumbers & Suppliers • Enhanced client satisfaction levels through the speed of process realized by their own end customers; • The automation of the way plumbers do business with Yarra Valley Water by reducing the time spent processing standard applications from ten days to less than 15 minutes; • Greater access geographically to services and property data via any one of ten plumbing suppliers who are linked to the Yarra Valley Water computer system;

	• The relevant approval, copy of the property sewerage plan and any other information such as special conditions of connection are automatically sent back electronically to the plumbing outlet without any manual intervention from Yarra Valley Water; • Paperwork is reduced and plumbers can pay on-the-spot for the service; • Plumbers realize the added convenience of picking up supplies while waiting for the application to processed; • EasyAccess is available to plumbers outside regular business hours: 7:00 am— 7:00 p.m. weekdays and 7:00 am— 5:00 p.m. weekends (soon to provide continuous weekend coverage) • Reduction in operating expense of an approximate $A3,000 per annum as a result of the reduced fee structure provided.
OPEN/coldplus	• Eliminated bulky paper documents and transportation • Real-time on-line data access • Contains records of current bills and billing history more than 7 years • Response time across 4.4M pages = 1.5 seconds • Tighter controls of financials and internal records
Implementation Approach	• Wang to assist in the development of a strategy to move from mainframe to client/server platform • Improve customer service • Integrate to all disparate systems • Encouraged high level of customer input to final solution • Pilot (customer) test by end users, real time changes prior to release • Internal pilot test by employees— complete three full application processes each— assisted in identifying bugs before release to customers • Software released to Yarra early January 1996
Implementation Team	• Round table approach • Members included staff, plumbing and lawyer customers, Wang and Staffware and other system solution providers • Team effort realized a high level of input • Client and customer expectations elevated

	through approach
User Education	• Minutes of training for lawyers, all mailed information package • Ongoing seminars for plumbers at the retail plumbing supply location, mailed information package, continuous market surveys at sites for system improvement plumbing supply retailers 2 days training two people minimum must include a manager • One day training for internal Yarra Water Valley employees, mandated the completion of every product application at least three times when training complete
System Documentation	• Product Functional Specification— Wang's development methodology, included formal documentation of system requirements. Document evolved through prototyping period. • Software Design Specification— containing internal system design. • "Hand Over" Guide— containing the site's operational characteristics for on-line operation support of the system. Yarra Valley Water help desk uses this document for training.
Major Integration with other vendors applications:	• The vendor technologies other than Wang's OPEN/image and OPEN/Coldplus technologies include: • Staffware: workflow software and for the front-end solution; • Oracle: database supplier; • Sun: Solaris hardware platform; • IBM: PC clients; • EDS: core billing and collection application; • TSW: asset management system; • Computron: financial system; • ConsulCAD: graphical information system.
Competitive Advantage	• Coverage to Plumbers over extensive geographical area thru plumbing supply retailers • Ability to submit applications while selecting supplies • All requests received via modem, mail or fax • Only water utility provider to have developed a revenue stream based on applications and requests for information

	• Requests fulfillment completed with no manual intervention by staff, staff morale improved and now have customer oriented focus • Ease of doing business attracting plumbers and lawyers to doing business with Yarra Valley Water • Same audience driving for the two other water utility providers to automate to the Yarra Valley Water model • Yarra Valley Water only water retailer of three to automate • Provides potential future revenue stream from other government utilities interested in solution • Measurable competitive advantage for the Victorian Government's measurement and ranking criteria • Level of customer satisfaction and service • Financial and bottom line performance • Environmental capabilities

Study: State of the Workflow Industry

Priscilla Emery, Sr. vice president,
Association for Information and Image Management International
AIIM

Executive Summary

The once popular notion of creating a paperless office is not the driving force behind adoption of workflow applications today. Improvements in efficiency and productivity as well as assistance with managing documentation and paper are the objectives of companies implementing or investigating workflow systems. Workflow products and services are becoming more established each year as more technologies mature to support them. The projected growth rates for these types of applications as determined from provider reported revenues and user estimates of expenses are estimated to be 55 percent per year, an increase from 50 percent for 1994 to 1995.

The workflow industry will continue to grow at a rate of over 50 percent per year through the end of the '90s. Most of this growth will come from new implementations by those companies that are currently considering workflow systems. Users who have systems in place and working, report (and hope) that their workflow expenses will go down. However, technological development makes more capable hardware and software available daily. This will result in systems in place today being significantly upgraded, and perhaps replaced, sooner than their owners forecast.

Twenty-one percent of those surveyed are dissatisfied with the systems currently available. This is somewhat natural for applications that are still relatively new and undergoing rapid change. The rapid change aspect of workflow products and services contributes to the most frequently mentioned obstacles to implementation, namely organization and user resistance and high learning curves. As is the case with all software and information systems implementations, gaining support for a new system and training the people who must use it are often the most important and difficult aspects of an implementation. Lack of flexibility to custom tailor workflow software to the specific needs and processes of a company also contributes to dissatisfaction.

Workflow Market Projections

The average 1994 revenue for both software and services providers, was approximately $10 million. Most providers are expecting their revenue to grow 50 percent or more each year through '96.

The rationality of this projection is verified by findings of the end-user portion of this survey that reports that 55 percent of the users across all industries are currently investigating workflow applications. Many of these companies will spend money on workflow applications for the first time in the coming 12 to 24 months. Much of the revenue in this market should come primarily from these new implementations.

New technologies and enhanced software will also add to the revenue growth from within the base of existing workflow users. Fifty-two percent of providers reporting in this survey plan to introduce significantly new products or services in the coming year. Existing users of workflow report their expenses will go down over the next few years. It is highly likely that existing systems users will find the advantages of planned upgrades in software and hardware too lucrative to pass up.

Based upon the combined data of the AIIM 1995 User and Provider surveys, we estimate the total market growth for workflow products and services to be above 50 percent per year over the next three to five years (on a 1994 market base of approximately $1.25 billion.) The combined market for workflow products and services will reach approximately $3 billion by early 1997.

Current and Expected Use of Workflow Products And Services

Presently, less than half of all companies in the AIIM survey base are involved with workflow software or services, however, 70 percent of them expect to use either or both in the future. We estimate that between 20 and 25 percent of all companies (including those not necessarily associated with AIIM) are currently involved in implementing or using workflow systems. Similarly, a large number will begin to use workflow systems in the near future.

Among companies surveyed, 19 percent use *both* workflow software and services, 13 percent use software, and three percent use services to support their workflow activity. In the future, five percent plan to use services alone and 23 percent will use software only. Approximately 15 percent of companies currently implement workflow applications outside the United States.

Today there are approximately 150 providers of workflow products and services. Of the 87 companies contacted for this survey, 22 percent provide software, 28 percent provide various services and 50 percent provide both workflow products and services. The companies responding to the survey represent an even mix of size, based upon revenues, from less than one million to more than $50 million per year. Smaller companies are more likely to be specialized and larger companies (over $50 million) more likely to offer a broader range of products and services.

STUDY: STATE OF THE WORKFLOW INDUSTRY

Workflow Usage by Respondents

Forty-three percent of all companies represented in this survey are using or implementing workflow systems. Another 44 percent are investigating the application of workflow offerings.

AIIM members comprise 64 percent of the respondent base for this survey. Fourteen percent of the AIIM members indicated they are not familiar with workflow software or services while 34 percent of AIIM members work in companies that use either workflow products, services, or both. AIIM members are more likely than non-members to use workflow services and software in combination in the future.

Seventy-five percent of the companies represented use workflow software for transaction oriented applications, 35 percent use the software for groupware purposes, 27 percent use it for handling company-wide forms and E-mail and 22 percent use the software for managing and documenting long-term projects and mission-critical documentation. In most cases workflow software is being used for more than one of the applications referenced.

Workflow providers report that most of their products and services support transaction oriented applications (71 percent). Collaborative/Groupware (41 percent), Publishing and project management (25 percent) and E-mail-based applications (17 percent) round out the remaining major applications.

Current workflow product activity is focused on workflow management software (59 percent), document management software (38 percent) and imaging (23 percent). Other specific offerings under development include COLD products, desktop applications, workflow engines, toolkits and OCR/ICR products. Just over half of these workflow product providers serve under 500 installations and the remainders serve more than 500 installations. Twenty percent of these companies report serving from 2,000 to over 10,000 implementations worldwide.

Workflow service providers report their services are currently focused mainly in consulting (62 percent) and systems integration (51 percent). Additionally, software development (40 percent) and training services (34 percent) round out the major categories of service offerings. More specialized services such as document conversion services and distribution are offered by 12 to 13 percent each. The number of users supported by these service providers is most likely to be in the 1,000 to over 10,000-user range.

The Technology Base

Slightly over 40 percent of companies currently using workflow software have Unix-based servers. For companies with revenue over $1 billion, 65 percent are using Unix server platforms, compared to approximately 40 percent of companies grossing less than $1 billion. Other commonly reported server platforms are: Windows (32 percent), OS/2 (14 percent), and Novell (13 percent). Proprietary mainframes are used as servers in 12 percent of companies.

The most prevalent client platform in use is Windows, with 65 percent of all companies using this system. Seventy-three percent of companies with revenues of over $1 billion use Windows compared to approximately 57 percent of companies grossing less than $1 billion. Beyond Windows, there are no other dominant client systems. DOS and OS/2 are each being used at client stations by 18 percent of companies, followed by Unix at 14 percent and Macintosh at 9 percent.

Fifty percent of server systems supported by workflow providers are based on Windows (25 percent) or Windows NT (25 percent). Forty-one percent are Unix-based. All other platforms, such as OS/2 (20 percent), DOS, Macintosh, proprietary and other operating systems are in smaller and often specialized concentrations.

Client platforms supported are dominated by Windows with 68 percent of providers reporting they support Windows client platforms and nine percent more support Windows NT. OS/2 is a distant second at 18 percent.

Industry Trends

There is little general consensus about which trends will affect the workflow industry over the next five years. Approximately 30 percent of users suggest that reengineering (BPR) will be a major influence on workflow applications, 26 percent mention a move towards an open architecture or integration of systems as a factor while 21 percent of respondents indicate the trend toward greater use of imaging will be a major influence on workflow systems.

Imaging and connectivity are seen as the technologies that will affect the greatest number of companies over the next five years. Forty percent of respondents indicate imaging is the key emerging technology. Thirty-five percent report connectivity technologies, such as servers and Local Area Network systems, are the technologies that will affect their company most in the future.

Workflow providers believe that technological advancements (43 percent) will be the largest influence on the industry over the next five years. The users questioned in this survey reported influences such as 'Re-engineering' or other business-related issues as being the most important factors in workflow systems change. The contrast of these findings highlights the need for providers to maintain an understanding of the perceptions and needs of the customer in product development. Other providers point to trends that users agree are important such as lower cost of implementation, downsizing / re-engineering and integration of imaging.

Providers view a wide array of emerging technologies as significant to the future of workflow. An almost even amount view improvements in software, networks, multi-tasking, groupware, and various aspects of multimedia as the most significant changing technologies. Hardware improvements, additional bandwidth in telecommunications, the on-line services phenomenon and improved imaging and OCR/ICR are included as well.

Providers and users alike expect easier integration and implementation of workflow systems and associated improvements in productivity to be the result of these technological developments. Providers hope for wider use of workflow software as a result. Elimination of paper is mentioned by five percent of the providers and was similarly downplayed by users.

Implications of Emerging Technologies

The users responding to this survey believe the implications of the emerging technologies mentioned to be:

1. Quicker and more complete access to information (24 percent)
2. More efficient processing (20 percent)
3. Expanded automation of processes accompanied by lower staff counts (14 percent)
4. Greater productivity and reduced paper usage (10 percent)

Most of the respondents (31 percent) who expect quicker access and greater amounts of information as a result of emerging technologies represent medium to large companies (those with revenue between $100 million to $1 billion).

While achieving a paperless office is still perceived as a goal by 60 percent of survey respondents, the ability of workflow software to reduce paper is rarely mentioned as a main advantage. Most companies with revenue below $100 million are processing fewer than 25 million pages of paper annually along with half of the firms with revenue between $ 100 million to $1 billion. As expected, a majority of companies having over $1 billion in revenue process the largest amounts of paper, beyond 88 million pages per year.

Importance of Imaging and Integration Capabilities

Imaging is certainly a technology of great importance to the workflow industry. Imaging technologies advance almost daily in capability. Thirty-three percent of workflow product providers report that imaging is a required element of their product offering.

Eighty-seven percent of providers indicate they integrate their offerings with the products and services of others. Users indicate integration with legacy and other systems is a major factor in their choice of workflow products and services.

Training Methods for Implementing Workflow

The training methods that workflow managers use or plan to employ to train their companies in the use of workflow tend to be more traditional and personal than automated:

- Group/classroom style (38 percent)
- Hands-on (15 percent)
- Personal instruction (10 percent)
- Computer-based (10 percent)
- Videos (7 percent)

Two factors reported as impediments to the implementation of workflow are the time required to train users and user resistance to change.

Satisfaction Level with Current Workflow Products and Services

Many users (41 percent) reveal general satisfaction with workflow systems with 12 percent being very satisfied. Over 20 percent are dissatisfied with their current software or services. Twenty-six percent could not assess their satisfaction level since workflow systems are still relatively new and are changing rapidly.

Survey respondents are in various stages of implementation. Satisfaction ratings, therefore, represent a wide cross section of perspectives on systems in various stages of maturity and change.

Workflow Software—Advantages

Most (64 percent) current or potential workflow users perceive the main advantages of workflow software as one or more of the following:

- Improving productivity
- Automating of processes
- Reducing processing time.

Sixteen percent believe that the main advantages of software are either to improve access and distribution of information or improve management and tracking capabilities. Improved customer service and cost savings are each seen as main advantages by approximately eight percent of respondents. Participants who represent companies with revenue of more than $ 100 million are most likely to view faster processing time as a main advantage of workflow software.

Workflow Software—Disadvantages

Twenty percent of respondents either did not recognize or did not articulate a major disadvantage in using workflow products. Fourteen percent believe that workflow software lacks flexibility (in programming or to modify flows) while 12 percent report that the learning curve for the user and training necessary to implement workflow products is the most significant disadvantage.

Workflow Software— Necessary Improvements

When respondents were asked about improvements they would like to see in workflow software, the majority did not specify any specific improvements. However, less than 10 percent believe firmly that no improvements are needed. The largest group, 32 percent, would like to see both flexibility and connectivity improved.

Workflow Services—Overall

The strong opinions regarding advantages realized from using workflow software are not evident in feelings about workflow services. Forty-three percent did not identify a main advantage in using workflow services. Expertise and knowledge of products is the only consistently mentioned advantage (19 percent) of using workflow services. Fifty-five percent of respondents did not specify a major disadvantage in using services either. The most frequently cited reservations about services were rooted in lacking trust of the supplier and lack of knowledge transfer from provider to the client company.

Providers, like users, report that rapid changes in technology are both an advantage and disadvantage. The main theme among providers regarding impediments to growth are market awareness issues such as user knowledge, skepticism, lack of acceptance and the need for more market recognition. There is a lucrative opportunity to develop effective marketing communications for these providers as well as effective training for providers and users.

Important Factors in Choosing A Workflow System

Ease of use and flexibility are the prevailing features considered during selection of a workflow product or service (44 percent). Other factors that would contribute to choosing one package over another are:

- Cost (27 percent)
- Compatibility with legacy systems (21 percent)
- Availability of product support (14 percent)
- Reputation (12 percent)
- Compatibility with legacy systems has greater significance for companies with over $1 billion in revenue.

The potential of a workflow system to improve efficiency or productivity is the main reason the majority of respondents (38 percent) would acquire an application. Saving money is another common goal (21 percent). For companies with revenue of over $1 billion, saving money takes on greater significance. Twelve percent of all respondents indicated that improved access would be the purpose. Thirteen percent of respondents did not identify any factors that would influence their decision to acquire a workflow system.

The Roles of Value-Added Resellers (VARs) and Systems Integrators

Ric Rhodes,
President, Document Imaging VAR Association (DIVA)

In today's world, value is the predominant idea, strategy, and tactical target that everyone in the world wants to offer. It is ironic that this old term (or philosophy) was attached to the Personal Computer resellers over a decade ago. The role of the "PC Reseller" declined, simply because PC delivery, combined with word processing, spreadsheets and hardware support agreements, turned into a base-line standard and was no longer perceived as valuable by corporations looking for new value in computing.

The PC resellers that survived this era where the ones who listened to their customers' needs and added the products, services and skill-sets customers demanded. As a reseller, if you *added value*, you survived. Only new Value-Added Resellers (VARs) and strongly established PC resellers emerged with value-added product and service offerings. These included offerings such as consulting, design, installation, training and support of everything that comprises a technology-based business solution— PCs, hosts, networks, telecommunications, and software applications— vertical and horizontal business systems

Before long the label of value-added reseller became synonymous with the role of the *systems integrator* responsible for converging multiple products and personnel together in deploying complex open-computing solutions. These disciplines are essential requirements for successful workflow and imaging implementations.

The true systems integrator is, and will be, the most frequently used and responsible delivery mechanism for providing the value corporations seek today and in the future. This is due to the fact that almost all software, hardware vendors, training/education specialists and consulting firms are teaming with or solely relying on the systems integrators worldwide to market, sell, deliver and support their specific product and/or service offerings.

Business visionaries understand the needs of their organizations and what will propel them into positive growth. A good systems integrator knows how to turn their visions into reality. Therefore, the secret of success is to select a systems integrator that will reduce your organization's risks, supply training and support the people responsible for changing the way your organization currently operates to your chosen new systems. The systems integrator complements your executive's macro vision and your consultant's slide presentation approach to business.

It is important to use focused expertise to solve a problem. This is similar to realizing your car brakes need to be repaired or replaced. You seek out the mechanic best qualified for the job, whether it is a Ford, Dodge, Mercedes or Volvo mechanic.

THE ROLES OF VARS AND SYSTEMS INTEGRATORS

Most common solutions involve multiple companies and corporate personnel that comprise the project team, including well-known consulting firms, both business and technical internal resources, corporate executive sponsors and lastly, but most importantly, the systems integrator.

The systems integrator should be the one resource willing and experienced in taking risks, eliminating finger-pointing during problematic situations and striving to ensure your projects' overall success.

A critical step—including the systems integrator during initial project planning

The Big Picture of where your organization is, where it needs to go and what value it provides are researched, analyzed and agreed upon at a macro level from management's perspective. This picturesque view of your future business approach is only valuable if there is a realistic plan in place that can achieve your organization's visions, strategies and objectives. Inevitably this vision will involve using the best available technology.

This is where selecting and contracting the right systems integrator will provide you with additional strength in accomplishing your success. Every astute business investigates today's technology, but it is difficult to comprehend how all these individual technologies work together or how much time and talent is required unless you've experienced these battles.

Perhaps bringing the necessary experts at this stage will incur an additional initial investment. This investment in expert resources, however, can keep your plans within the realms of success and equate to exponential dollar savings by leveraging the diverse experiences and technological know-how of the systems integrator.

Today's corporations tend to create internal focus groups to investigate imaging and workflow application functionality and the value these potential business solutions offers the organization. The formation of internal focus groups is definitely not cost-free. They can result in an expensive venture with corporate personnel moving without much experience or knowledge into uncharted waters. To avoid delays and non-expert advice, the logical solution is to out-source experienced technological resources during the project planning process. You need resources that are fluent with workflow and imaging technologies and business practices.

This initial investment will highlight possible shortcomings that may evidence later and it will mitigate potentially detrimental results that could involve major strategic and tactical reorganization. The big picture investigation by the systems integrator involves which hardware and software products are your best solution, what resources are required for solid implementation and what continuing support should be included in training and future growth.

VARs' abilities in pulling it all together

A good systems integrator specializes in knowing or quickly researching which software and hardware products work well together. They also have the technological depth to sustain the huge effort required to integrate these technologies to produce effective information management systems. Imaging and workflow projects involve managing and routing all data types— paper, pictures, electronic files, video, voice and host computer reports. In essence, the challenge requires artful packaging and presentation of the critical organizational data required by your organization. But, just as the foundation of a strong building is the heart of the structure, so is the network infrastructure of information systems. The systems integrator also analyzes your current infrastructure to ensure it can carry the demands of imaging and workflow applications.

After collaboration with your systems integrator, you will have identified your winning imaging and workflow solution. You will then realize that this solution involves a multitude of major components, such as new and old databases, GUI development, the Internet, imaging hardware and software, workflow implementation, administrative tools and other assorted software, optical libraries, server hardware, networking technologies, etc.

Pulling these disparate technologies together is where your systems integrator shines. The systems integrator is the soldier willing to participate in this tough battle of fighting for success and returned value. The big question is whether your particular solution requires explicit focused business or technology expertise. It is imperative to combine your technology and business efforts to gain the full benefits of workflow and imaging solutions.

The systems integrator you need is one who employs business analysts, programmers, network engineers and technical people, all of whom will work directly with your own business and technical personnel. The most critical components of deploying a successful imaging and workflow application are the abilities to define and develop the structure of training and education curricula. These programs should include all personnel affected by the application. There will be multiple levels and types of curricula from awareness training for executive management, to procedural orientation and technical education for systems development and administration personnel. This training will ease your organization's transition into the new ways of working with information.

Another advantage of using a systems integrator is their fountain of industry knowledge. It's imperative to understand the state of the art concerning new product development both today and six to 12 months ahead. Systems integrators are usually involved in multiple vendor beta programs that allow them to experience and analyze new features and functionality. This hands-on approach can eliminate custom-developing a needed feature or function that may be available in near-future product releases. Integrators also keep current on pricing related issues concerning product bundling, new software and hardware releases.

This knowledge enables your systems integrator to provide high value during the design and planning of future project phases, including application enhancements and additional users. Your organization can rely on the systems integrator for product pricing, maintenance and quoting costs for professional resources to take your business solution to the next level.

These developments, enhancements and expansion should be taken into consideration when measuring the success of your project. This will involve polling the users and project members on the usefulness of the existing system, what benefits the application provides today and what they foresee it providing in the future.

With this evaluation data in hand, the project team should analyze the agreed-upon design features and functionality. This may involve selecting additional tools, adding functionality to current tools or planning the next phase of the project. Remember that business requirements constantly grow and change and project phases should be designed to complement these changes. Define new or expanding business objectives and work with the systems integrator to establish continuing solution-oriented growth.

The project should never sleep. Even when you have measurable success the project doesn't end. On the contrary, a successful project that has been justified can now grow either vertically or horizontally throughout the organization. Helping other divisions and expanding the user-base of the system is best accomplished during this period.

This growth will involve similar issues concerning what technology, business processes or people within the organization are in need of improvements. It's a revolving door that should never remain idle. This is especially true if your organization has a sworn mission to strive for continuing improvements and excellence.

The project team may change, involving different business personnel, maybe different technical resources, but you can always rely on the expertise and experience of a systems integrator to supply your organization with continuity and the knowledge your organization requires to complete complicated technology-based business solutions.

There are many types of systems integrators, and they come in all flavors and sizes. Look for specific qualities in your systems integrator, such as certain application and business expertise. Always seek an integrator who maintains hands-on technological knowledge and proven implementation excellence. There has been a lot of confusion concerning the roles of consulting firms and true systems integrators. It's only in the last five years that consulting firms have started rolling up their sleeves for implementation of everything from networks to client/server development, but typically they require seven digit fees on their projects.

True systems integrators have never hesitated to roll up their sleeves to get the job done, whatever it may take. They are much closer to the technology and their referenced implementations should depict their experience and credibility, not to mention generally lower fees.

In the foregoing pages of this book, you will have noticed the essential roles of the integrator and VAR in the excellence of various workflow and imaging implementations. Study how they managed and integrated these projects and discover parallels to your own plans. Right here is a good place to start your VAR and integrator selection process.

Pitfalls in the Strategic Deployment of Process Improvement and Management

Geary A. Rummler and Gordon Sellers
the Rummler-Brache Group

The Organization as a System

The successful implementation of process improvement and process management is the most significant management trend impacting business operations to come along in the last 50 years. As with all trends, there has been an onslaught of management paradigms. Programs such as *Total Quality Management, Benchmarking, Reengineering, Just-in-time*, etc. are endless in their iteration. Although all valuable additions to management's arsenal of tools, none of these programs, individually or in combination can save senior managers from expending tremendous effort formulating and implementing the plan-de-jour.

There is no question that the notions embodied in these programs fit into the process manager's toolkit. In fact, the singular commonality among them is their focus on process. However, to be successful, focusing on a process is not enough. We assert that in the future successful companies will be best described as adaptive and as ones that manage their organization as a system of inter-related and interdependent processes.

Organizations that are being managed as a system can be characterized by the following:

1. the continuous, systematic monitoring or external events and internal performance
2. alignment of the organization, process and job levels within the company
3. fast, rational, coordinated responses to external events, performance issues and trends
4. the continuous improvement of the management system to improve speed and effectiveness of adaptability
5. strong leadership
6. an infrastructure designed to support a performance measurement and management system
7. a behavioral shift from traditional management thinking and approaches to those that support new process centered management approaches

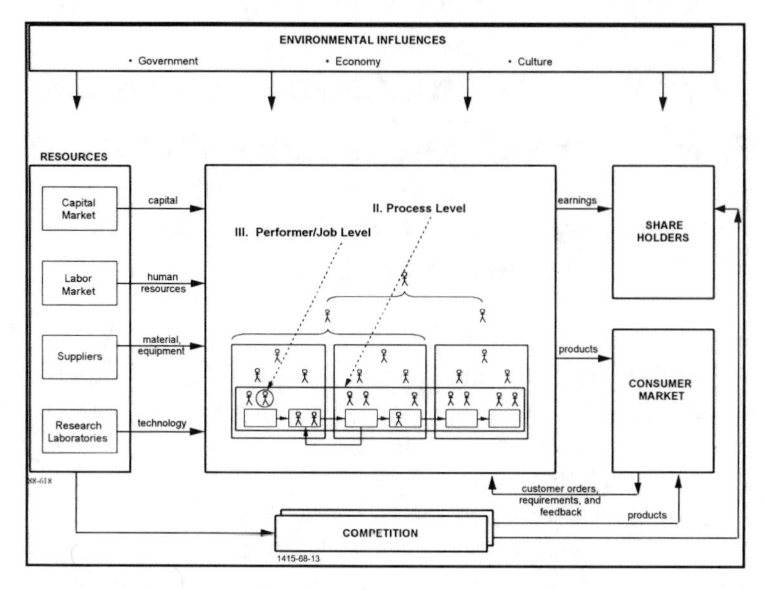

The systems view of the organization

In the following pages we identify and categorize the most common pitfalls that our experience has shown lead to unsuccessful or incomplete deployment of the process improvement or process management approach to running your business as a system.

Pitfalls in Implementing Process Improvement and Process Management

Our experience in hundreds on engagements with clients has shown that there are really only two categories of pitfalls awaiting the unwary process manager. The first level is those pitfalls related to the overall strategy driving the process improvement initiative. You will find them in organizations initiating a major process improvement effort involving multiple processes. The second set of pitfalls relates directly to the act of improving any process.

The Major Pitfalls in Initiating a Major Process Improvement and Process Management Effort

This list represents the top pitfalls for implementing major process improvement and management efforts. These pitfalls are typical of troubles we have seen on major engagements where more than one process is being improved at the same time.

1. Failure to have, communicate, and reinforce a sound business case (including a sense of urgency) for why the organization is undertaking the effort.

As obvious as this may seem it is consistently trouble for many organizations. All to often in their search to leave an indelible mark on the organization senior executives instruct their staff to "go out and do it." We routinely find this to be the result of somebody high up reading the latest business journal article on the hottest new management approach. The trouble is that the hot new management approach does not tie to a real world business case and the team is destined to stumble over themselves "doing it."

What tends to happen is that the management team (one level down from the boss) buy into the executives' vision, with all the VPs vying for the boss' approval because they are "doing it" better, while in reality the folks out on the shop floor or in the field see no real change in the way things get done.

To have a sound business case for change you must have a clear tie to the organization's mission or more particularly the organization's strategy. This usually is a statement of some sort of financial, economic or market related goal (e.g. we will reach $25 million in revenue by the year 2000). If no tie to the organization's strategy exists and if it is not communicated effectively, the process improvement work will not be focused on supporting and attaining the corporate objectives.

2. Mis-conceptualizing, identifying, and labeling processes.

A process is a set of steps/activities for converting specific inputs into specific outputs. In the excitement of getting "process focused", frequently everything becomes a "process", including functions and issues. This is a particularly deadly pitfall from which it may be impossible to recover. Many Business Process Improvement (BPI) efforts are fundamentally flawed from the outset because of the way processes were initially defined. For a real impact on the performance of the business the selected processes must by tied to a critical business issue that needs resolution.

3. Start at the bottom of the organization (empower the people to fix their processes) rather than at the top.

You should start at the bottom because:

a) Process performance requirements should come from the customer or the business objectives of the organization. This critical direction usually resides at the top of the organization.
b) Significant change must start with improvement of the *significant cross-functional/organizational* processes that touch the customer or critical business issues (e.g., order fulfillment, product development, business generation, planning.) Almost all process improvement work within functions should be initiated *after* the cross-functional process work is started, and be in support of those processes and that work.

4. An "everybody out for a pass" type effort.

This, too, is deadly. The idea seems to be that *process improvement* is good, and therefore everybody should "go out into the wilderness" and do it. ("It" being form teams, pick a process— almost any process-and try to improve it.) The reality is that:

a) Not all processes need to be improved. Many should be done away with, in fact, and that is something that few job incumbents will have the courage/stupidity to do.
b) Redesigning processes effectively and efficiently is hard work (per the process improvement pitfalls cited below) and time-consuming, and not everybody has the time, skill, and inclination to do a good job. Beyond the *redesign* of a process, the *implementation* consumes major resources and requires major amounts of management support.

Recommendation: Launch BPI teams carefully to address specific, serious business issues. These efforts consume major resources and should be launched only when a good business case can be made. We can guarantee that a BPI project, for which there is NOT a ***strong*** *business case, will not have its findings implemented and will deliver no results.*

5. Getting the awareness training out ahead of the real process improvement work. (Perhaps this is a corollary of #4).

Frequently, in an effort to show the organization that management is serious about BPI, they initiate a major BPI introductory training program and "sheep-dip" hundreds of employees. Usually this is done before there is really a plan for how the company intends to roll-out BPI and months before any of the trainees are likely to be personally, seriously engaged in a BPI effort. Or, the trainees decide that this is good stuff and they begin to experiment with it in their job. In either case, the results are going to be less than optimal.

Recommendation: Do initial training on a JIT, as needed, basis. After you can report results in strategic pockets of the organization, then consider general training on the wonders of BPI for broader audiences.

6. Focusing on Business Process IMPROVEMENT while not address the critical issue of Business Process MANAGEMENT.

If there is not some corporate level management system in place, whereby senior managers are continually asking about process performance results, there will be no sustained process improvement. It is essential to link the process measures to the "enterprise measures" at the top of the house and to monitor how the processes are contributing (or not) to the overall performance of the business.

The Major Pitfalls when improving a Process.

This list applies to the improvement of any single process.

1. Process improvement effort NOT tied to the organization's strategy.

For a process to be accurately designed and successfully implemented, the process improvement/redesign must be linked to a Critical Business Issue. Without this link, there will be no performance targets for redesigning the process and no way to measure improvement. Also, without a compelling business reason to improve the process, the process improvements will never be implemented.

(Corollary to #1: Teams starting out to map *processes rather than to* improve *processes. No process mapping efforts should be started unless it is clear that the objective is to improve the performance of that process. Such efforts will end in countless hours of meetings and flowcharting, and no results and very frustrated and disillusioned "team" members.)*

2. Not having a clear charter, objectives, measures, and sponsorship (i.e., process owner).

The process owner can not be an employee from the rank and file fitted with the impossible task of swimming upstream. Rather the process owner must be a member of the management team that has the most stakes in the process being improved. He/she should be the champion of this process and of the people who perform the activities that make up the process. Not only will this individual have to manage the individuals within the process— which may come from disparate functional groups, but he/she will have to manage the intersections between functional groups through which the processes pass through; the white space on the organization chart where finger pointing and fumbled handoffs can occur. Additionally, in the process-managed organization, this person would be accountable for the process results.

3. Nibbling *at process improvement, 2-4 hours at a time, for months, rather than a focused effort in a short time span.*

The "nibbling" approach almost always ends in project entropy and disillusioned team members. This is serious business. For any true change to take place a serious effort has to be made. True effort takes time and resources. If the time and resources are not available to focus on the process improvement effort, than it is better not to make the effort at all as the results will be wholly disappointing.

4 Lack of clear accountability for results.

What is required is a Process Owner accountable for improved process performance, a Steering Team (or equivalent) who is accountable for implementation, and a Design Team accountable for the design of an appropriate "should" process.

5. No agreed upon approach or methodology for analyzing and improving processes.

The team makes it up as they go along. Or each "facilitator" tries to follow an approach they read about in a book or magazine. If these projects ever succeed in reaching a viable process, they won't really fit together because everybody was using a different approach, language, etc. For an overall, coordinated, effort it is crucial that teams have a common approach to BPI. At Rummler-Brache we are willing to teach them the time-tested methodology RBG PIP methodology. The methodology is robust enough and flexible enough to accommodate modifications by various teams, so they won't feel like they are in a methodology straight jacket.

6. A tendency to over-analyze the existing process.

Taking weeks, rather than the day or two. This tendency is usually because the teams don't have a sound methodology and are unsure of how to approach the design of a "should" process. The result is everybody hiding-out in the "is" analysis.

7. No "Should" Design Specifications (A subtle, but key point..)

Without this, the teams will seldom design a "should" process that achieves the desired performance outputs.

8. Failure to drive the redesign down to the Performer Level, redesigning key jobs and their performance support systems.

There is no point in redesigning processes if the jobs you ask the employees to do not support the goals of the process and reward them for performing within the perimeters of the process management system.

9. No methodology for implementation.

As with the major pitfalls above, the team makes it up as they go along.

10. Failing to understand that improved processes need to be managed.

They do not manage themselves. Failure to put in a place a *Process Performance Measurement and Management System* for all key processes is to pour all the process improvement effort down the drain. Avoiding these pitfalls is the responsibility of the management. It is easy to fall into the traps that lay in wait for the unwary process manager. Real, sustainable and measurable process improvement and process management is difficult and challenging things. To succeed, process improvement and management must be a corporate imperative that is driven from the top down. It must be ingrained into the thinking of every manager and every employee and it has to be tied to the strategy of the organization.

As a manager you must be prepared to manage the process, for if you are not prepared to continuously manage processes, you should not be surprised if you are asked to continuously fund large scale, ad hoc process improvement projects.

The Seven Traits of the Smart Company

Excerpted from the book, Smart Companies—Smart Tools, By Thomas M. Koulopoulos1997. Van Nostrand Reinhold New York

What will tomorrow's companies look like? After rampant innovation in technologies such as workflow, Intranets, and the desktop the question has nearly become a parlor game— it's anyone's guess, none is wrong, none is right. Yet there are critical trends that are evolving and shaping the organization of the future and turning companies into what I term *Smart Companies using Smart Tools.* The traits these companies share are similar, although their cultures and industries may be radically different. The key in the case of every smart company is that it relies on a foundation of technology to achieve its critical objectives, and that these technologies increase human dignity rather than denigrate it.

How do Smart Companies get smart? They begin when they realize that it is impossible to divorce the communications or the fragmentation of an organization from its technology.

One of the risks that has accompanied the trend towards more distributed organizations is that of highly distributed computing environments and the fragmentation of processes. The democracy that epitomizes desktop and PC computing also accounts for isolation and miscommunication among different functions within an organization. In many ways we could say that the division-of-labor popularized in the early part of this century by Sloan and his fellows has turned into a division-of-computing.

The Tower of Babel is said to have been created by legions of people who all spoke the same language. Their ability to communicate in a single tongue gave them the basic tools with which they attempted to build a structure that would touch the Heavens— spanning the chasm from mortals to the divine. This anathema to God did not go unpunished by the creator who in his rage destroyed the tower and scattered the legions across the world into separate and isolated geographies where they were each forced to use differing tongues to communicate among their individual tribes. So was born the nation state, according to the biblical prophets, and thereafter mankind was relegated to miscommunication and misunderstanding, which prevented them from ever again easily sharing a glorious common vision.

Even the prophets, it seems, understood the value of communication as the foundation for the greatest endeavors of mankind. In fact we could even go so far as to say the it is communication which separates the gods from the mortals— and the gods are not quite ready to consider us equals on this score.

Another reason for this fragmentation is the plethora of computer hardware and software vendors that has sprung up in the last 30 years. These vendors have relied on differentiation at all costs to rise above the noise and set themselves up as the standard. In the process, proprietary computer systems have become the norm. The running joke among both vendors and users has always been that there is no lack of standards in the computer industry, in fact there are plenty to pick from.

The marketing strategy of technology vendors that lead to this diversity was valid but short sighted. The greater need for any information system can not be subjugated to technology, and that is the competitive requirement for a networked economy, where companies can easily communicate and transact with employees, suppliers and customers. This in turn requires a networked and interoperable computing infrastructure that will act as the basis for collaboration and communication among the participants of that new economy.

Smart companies are turning this tide of proprietary platforms and fragmented organizations by relying on technologies such as workflow and applications of these technologies which break down inter/intra-company boundaries. By doing this they are also establishing a new competitive benchmark for innovation, response times and customer service.

To do this smart companies share some basic philosophies that support their efforts to break down these barriers. The philosophies are both technology-based and cultural, but specific smart tools, which I will talk about later in the book, ultimately support them all.

All of these philosophies focus on increasing communication and collaboration between people and processes. Only through increased intimacy and understanding of the processes that we inhabit can smart tools help us to dramatically improve our work environments.

The most prevalent smart company philosophies are:

1. An integrated rhythm of work
2. Fostering a high degree of process intimacy among employees
3. The use of asynchronous communications to bridge time and geography
4. Applying technologies to leverage rather than eliminate people
5. A strong emphasis on Return-on-Time as the principal success metric
6. An extended enterprise that encourages non-traditional employment
7. Heavy technology investment in the organizations Touch-Points

An Integrated Rhythm of Work

As technology quickens the pace of work, we all feverishly try to keep up with the metronome. Yet most of us are continuously falling behind. The reason may very well be that we have become so much better at transferring work to any location than we have been in coordinating its performance. Virtual workplaces infiltrate our lives at all times and places, often wreaking virtual havoc. Smart companies invest heavily in tools that ease this burden by helping employees control and manage their work, in the face of ever increasing and portable work.

It wasn't that long ago when we associated a distinct place and time with work, family and leisure. Work was done in the office during the weekdays from 8-5; evenings and weekends were for family at home; leisure fell somewhere in between the routine of work and family. As our society and its technologies have changed, however, the lines of demarcation between these three areas have become increasingly more vague; some would say more complex. We weave in and out of the three regularly.

For better or worse, our lives have become entangled webs of connections, telecommunications and networks. Advertisements promise that *you will* be connected at all times, and from all places, to those you love. But the connections work in two directions. Yes, we have the ability to connect with our families and loved ones around the globe as well being connected to our work. It is these connections that are increasingly becoming the backbone of an around-the-clock work ethic. Boundary-less work has become expected and increasingly mandated. For instance, on a recent plane trip, I sat next to a lawyer who told me how his firm required by contract that he be not only available at all hours, but carry a modem-enabled laptop, cell phone, and pager as well.

Vacation? What's that? A change of scenery? As Alvin Tofler originally said, "Any-place, anytime." If there is a phone line nearby, work awaits.

Amid all this, we hear in the hallways of global corporations the familiar themes of virtual enterprises and a new work force of free agents. Yet all of these connections do little to add quality or even productivity to the enterprise we work for if there is not a mechanism by which to coordinate their myriad tentacles in such a way that they can become part of a sane work ethic. Connections to work can become a maddening treadmill if they are not managed against the priorities of our lives. To paraphrase a popular cartoon, the good news is that companies are constantly creating new jobs; the bad news is that these new jobs are just being piled up on top of the old jobs, already stacked two and three deep for most workers.

The problem is simply that managing the priorities of ever-increasing workloads has become the property of nearly every professional. Our desktops lie under the clutter of an information deluge that is only getting worse. We are our own administrators, support staff, managers, and workers. With 10,000,000 fewer secretaries since 1989 in the US work force alone, it's easy to see how this is a problem gaining momentum.

Is this a technology problem? Can smart tools help? It seems to be more a problem for the social rather than the computer scientists! Technology is, in fact, the only way to solve the problem that technology has created. Specifically, that technology mechanism is one of the most important smart tools— workflow. As a discipline, technology and management tool, workflow provides the foundation for far-flung change in the way we work. More importantly, it allows this change to occur in ways that increase the general quality and rhythm of our lives.

How? Workflow automates the steps that go into a business process by capturing the rules, routing instructions, and roles required to perform work. Once a workflow is defined, the work progresses automatically from one task to another, avoiding the inherent delays associated with manual workflow.

If we shift our focus from the delivery of work, to the coordination of work, technologies like workflow may actually make our lives more pleasant by proving that work is best accomplished when it is integrated intimately with our lives.

If all of this sounds a bit too Orwellian for your taste, consider how basic technologies like the fax machine, cellular technology, pagers, and e-mail have already altered the rhythm of your life. Like it or not you are accessible. The question is not how to break away, but rather how better to integrate and manage the accessibility. Smart companies realize this and create environments for work that provide tools for coordination and work management, rather than burying their workers in the complexity of the connections to their work.

A high degree of process intimacy

One of the most basic problems in today's highly specialized work forces is that of process intimacy. It's actually better phrased as a lack of intimacy. Simply put, increasing task specialization makes each person in a business process isolated from the others. The isolation may be not only in the form of *space* or *distance*, it can be *time*. But the result is the same.

When two people are separated by several days in performing their respective tasks on a particular value chain of activities, the reality is that they are less likely to understand the impact of one another's work than if they were separated by hours or minutes. Why? In a word— *iteration*. The longer it takes to iterate a task, the less likely it is to be iterated. In other words, if it takes you two days to get a response to a question as opposed to two minutes you are less likely to ask the question. By the same token, you are less likely to try to understand or change a process if you are removed from its components by significant intervals of time.

Smart companies close the time intervals by eliminating the inherent transfer times in routing information and work from one person to another. Interestingly, enough, however, this is not a problem that can be solved by a communications network alone, no more so than it can be solved by the migration from paper to electrons. Electrons may travel at the speed of light but work does not. If you don't believe that, when did you last read your e-mail? If it was two hours ago, then you could say that the e-mail messages waiting for you have taken at least that long to get from their sender to you, although the actual message traveled at the speed of light and probably took mere nanoseconds to be delivered. The same is true for work of any sort being delivered electronically.

Smart companies go beyond networking alone by using sophisticated mechanisms to route the work and deliver it in the fastest possible period of time. This may include using roles to route the work instead of a specific person. In this case, if one person is not available to act on the work, the workflow will find someone else with similar qualifications to perform the work. The key is getting the work done not just transferring the information. This is often referred to as a pull model, as opposed to a push model of work.

In the pull model we don't have to hunt down the person, to get the work done, instead we go after the role. Without smart tools technologies such as e-mail become a form of technology deluge. Ironically the push model works best for employees who perform relatively poorly and worst for those employees who are most productive and responsive. The reason is simply that work tends to congregate around productive people, creating an incredible imbalance. Knowledge is punishment in this model since the rewards for work well done are an ever-increasing workload. Pull models tend to distribute work more equitably among all workers defined by a particular role. Additionally, as you will see with the case study of Momentum Life, the pull model can be used to create more equitable forms of compensation for workers.

Smart companies believe strongly in this equity and create work environments that recognize it.

Established Paths for Asynchronous Collaboration

Another profound effect of the modern organization has to do with the nature of global communications. At the heart of this is the concept of asynchronous communication; a means of enabling communication that allows business processes to bridge time and distance. The concept of asynchronous communication is simple yet powerful. Standard interpersonal communications occur in a synchronous mode. That is, when two people conduct a discussion, they are able to communicate at the same time and possibly in the same place. A benefit often attributed to synchronous communication is the ability to address issues as they arise without delay. Asynchronous communications, on the other hand, occur in series without interaction and interruption. Although contemporary forms of asynchronous communications come to mind, such as e-mail and voice-mail, asynchronous communication is as old as mankind. Etchings on cave walls, and every form of written communication since that represents an asynchronous communication mode.

In a global context, however, asynchronous communications are often associated with delays since each party is always waiting for the other before continuing the communication. This occurs with groupware when a task involving several messages (communications) requires that a response correspond to each message. This can be especially problematic if the parties communicating are separated by several time zones. The following example shows the effect this will have on a simple global task involving several communications.

When Boston sends a message to Hong Kong at 9:00 AM (10:00 PM Hong Kong), Hong Kong will not receive the message until the beginning of their day (10:00 PM Boston). Although Hong Kong may work all day on the task and make significant progress, Boston will not be aware of the progress until the task is complete and they receive a communication from Hong Kong. If Hong Kong sends this message on the day following their receipt from Boston, the total delay from Boston's perspective is at least two days. The total task time, however, is only half of the total elapsed business cycle time. That means that our business process is only 50 percent efficient.

That may seem like an insurmountable problem. After all, smart companies cannot alter the arrow of time. But they can eliminate the fundamental obstacle in synchronous communication, the concurrence of human communication. With an automated workflow system, the asynchronous communication can continue even though the people are not synchronized. In our example with Boston and Hong Kong, Boston would be able to communicate with the task or the process at any time. I refer to this as P2P communication, where the Ps stand for Process or Person. This allows for three separate interactions: Person to Person; Person to Process or Process to Process.

This is a fundamental change in the nature of communication, since you are no longer communicating with people, but with the process itself. It should be noted here that communication in our definition requires that there be real-time interaction between the two parties. (The parties, in this case, could be any combination of humans and process.) Without an automated workflow system, this can only be accomplished if two individuals are both available at the same time. Workflow captures the process in such a way that this communication can continue at any time. This fundamental shift from person to process is one of the most significant aspects of smart companies, and may similarly be one of the most profound advances in global process automation.

Applying technology to leverage rather then eliminate people

Far too often we look at the technology as a replacement for people. Although it is clear that there are many tasks in any organization that can be automated, smart companies view technology as a means of augmenting people. The mania to focus on head count reduction appeal to a short sighted mindset that lacks the creativity to consider how technology can free people and organizations to explore new avenues of opportunity, new products, and increased leverage of its human resource.

Smart companies do not rank well when compared on the basis of people eliminated from the application of technology. They do, however, win the prize for number of jobs obsoleted by technology. The difference? Job definitions should, and will, change in competitive markets, but people can move into new roles and new jobs as the old ones are left behind.

In smart companies technology is used as a method of augmentation to best leverage the people, skills and products of the enterprise.

A strong emphasis on Return-on-Time as the principal success metric

Everyone talks about return on investment, smart companies talk about return-on-time. Ultimately all costs hinge on time. One way to look at a business process is to consider how much value is being returned for a given interval of time invested. (This is specifically what we will do in the section on Time-based Analysis.) In the case of an individual, that time is measured in a literal sense— the number of heartbeats invested. You are trading yourself for a given deliverable. The irony is that continuous work results in lower return on time than intermittent work. In other words, if the work is spread out over time and performed in increments, there is less actual time spent waiting, queuing and correcting work.

I saw this demonstrated vividly in an approval cycle at a large manufacturing client. Vouchers for parts ordered by shop floor machinists required approval by accounting. In a typically regimented fashion, accounting would periodically review orders twice weekly, then send the order to the originator for approval. Since machinists were busy working other machines during the day, the orders were often delayed by at least another day as the machinists reviewed their in-boxes at the end of each day. In total as many as four days would go by for single approval iteration. In many cases, more than one iteration was required, leading to a worst case scenario of two weeks to approve a part. An automated workflow allowed accounting to immediately route the approval request to the shop floor where machinists could periodically check any one of several networked computers during the day. The approvals were now integrated with the work rather than an impediment to it.

Return-on-time is a measure of how well an organization balances its investments in smart tools with its critical success factors. If filling customer orders in less than 24 hours is essential for success in your industry then your investment in timely order fulfillment is easy to assess. It becomes what I term an *escape velocity* question. In other words, no matter how much you invest if you do not get the fulfillment process in at under 24 hours you have failed— 36 hours, 28 hours, 25.5 hours are all unacceptable no matter what it costs.

An Extended Enterprise that encourages non traditional employment

The evolution of virtual workgroups has been fueled by at least three major trends: increased availability of white-collar unemployed (a.k.a. consultants who do not work in the confines of traditional office environments); cheap portable and home office computing; fast reliable communications. As a result more companies are opting for non-traditional work environments which include a significant number of workers outside of the corporate office. (I will take a closer look at this phenomenon in the last chapter of the book). Smart companies have been especially quick to embrace the opportunity presented by these

Although telecommuting and work-by-wire have quickly caught on they are fraught with coordination and collaboration problems which are causing some companies to look at ways to minimize the technology investment in remote set-ups, while still realizing many of the benefits of an extended workforce. These are solutions that beg for smart tools. Companies who master the power of a remote workforce are not simply using work-by-wire, however. In many cases, such as Nordic Track, the key is using core personnel in the most effective way while also relying on non traditional workers to shoulder much of the responsibility for certain core functions.

In a dramatic example of the virtual enterprise in action, Nordic Track's new product innovations stem primarily from an ever-growing network of independent inventors. Amazingly, the company solicits these inventors through classified ads in back of magazines such as *Popular Science* or *Design News*.

As a result of its unique R&D method, Nordic Track receives about 100 invention submissions a week, ranging from sketches on cocktail napkins to complete CAD files and multimedia presentations. According to Wes Cutter, Nordic Track's Vice President of Product Planning, "The inventions we get are just as diverse, spanning everything from pogo stick-driven bicycles and perpetual motion machines to ideas that resulted in some of our most successful products, like the Flex Gold isokinetic strength training system or the downhill skier machine."

To support its growing inventors' network (and to ensure that a good idea doesn't get lost under a mountain of unprocessed submissions), Nordic Track established a policy to review and respond to every inventor within 10 days. But with its existing manual procedures, achieving this 10-day response time would have required legions of additional staff just to manage and route all the paperwork.

"Our only solution was to automate the business processes in our product development group," said Dave Janiszewski, Manager of Office Automation Services at Nordic Track.

Smart Company "Touch-Points"

There seem to be limitless possibilities in any organization for the application of technology, making a frequent question of many companies. Where to begin? Smart companies typically pick from three priority areas that involve interactions with customer, suppliers, or employees. I refer to these as a company's touch-points, since they touch the most important point of interaction for most organizations.

Each touch-point represents a significant area of potential process or quality improvement, and competitive advantage. Most importantly, touch-points represent areas where human interaction is often at its most intense.

Touch-points can be regarded as the periphery of an enterprise's central nervous system. Like the human neurology, the extremities define the efficiency of our interaction with the world around us. Dexterity, mobility, and adaptability depend primarily on the nimbleness of our peripheral nervous system; fingers, toes, hands, feet, arms, and legs define how well we can react to events around us. Granted, the genesis of all reaction is in the brain, but actions are not expressed here. In this same way an enterprise may have outstanding strategies, plans, and tactics, but they must be enabled in the actions taken to satisfy customers, educate and leverage workers, and negotiate and partner with suppliers.

If the fundamental premise of smart tools is the liberation of human potential, and its application to the areas of greatest value, then smart tools must increase human interaction at these touch-points, while streamlining the tiresome tasks that otherwise consume workers, customers, and suppliers.

A slight application of the right technology in these areas can have extraordinary impact on a company's processes.

It could be argued the most important touch point is that of company/customer, and this often is reduced to the contact between a customer and a customer service representative. But what does a company do if the customer service staff turnover is high and training opportunities are at a minimum? It's easy to take an academic stance and say that they simply need to retain employees, but there are simply far too many customer service jobs that act as stepping-stones to other positions. In addition customer service has always been a difficult position to maintain because the accolades are often hidden. When customers are happy they rarely sing the praises of the customer service, but when they are irate customer service is the first to be noticed. Creating an environment that both rewards and recognizes customer service, while facing the reality of turnover is momentous challenge. The solution has to combine Smart Tools and smart management.

Start Smart

Companies who implement and foster these seven traits will achieve competitive advantage and increased productivity to be sure. More importantly they will achieve greater prosperity among workers, customers, and owners. Remember, however, that smart tools are just that— tools. The way they are used will change from one organization to another. Methodologies are not meant to be cults, they are not dogma— they are simply guidelines. Be leery of anyone who tells you otherwise. An education awaits you, and the only way to begin is to start today. Start large, start small, but start smart and don't expect ever to stop.

- Violino, Bob. Bringing Harmony to Business Systems. InformationWeek. 547 (October 2, 1995).

- Escape Velocity is the speed which an object, any object, must attain in order to escape from the earth's gravitational pull and enter Earth orbit. Even a slightly lesser speed will simply not work. In the same light there are many success factors that are measured in these terms since no less than a specific measure will attain customer satisfaction, or competitive parity. The amount invested has nothing to do with success if the success factor is not attained.

Contact Information

Arrow Trucking

North America Excellence Awards: Imaging, Finalist

Name: Don Battle
Title: Director of MIS
Phone: 918-446-1441
FAX: 918-445-5714
Mailing Address: 4230 S Elwood
Tulsa, OK 74107

Lanier Worldwide, Inc.

Name: Emmy Brock
Title: Marketing Specialist
Phone: 770-621-1472
FAX: 770-621-1062
Mailing Address: 2300 Parklake Drive, NE
Atlanta, GA 30345-2905
email address: ebrock@lanier.com

J.D. Young Company, Inc. (Integrator)

Name: Robert Stuart, Jr.
Phone: 918-582-9955
FAX: 918-582-2085
Mailing Address: 116 West 3rd St
Tulsa, OK 74103

Bank of America

Asia Excellence Award: Workflow and Imaging, Silver

Department/Division: Bank of America, Asia Division
Name: Doug Switzer
Title: Senior Vice President,
Global Wholesale Operations
Phone: 415-622-9947
FAX: 415-953-6218
Mailing Address: 555 California St, 42nd Floor
MS 8676
San Francisco, CA 94104

FileNet Corp.

Name: Colleen Edwards

Title: Public Relations
Phone: 714-966-3400
FAX: 714-966-3490
Mailing Address: 3565 Harbor Blvd
Costa Mesa, CA 92626
email address: cedwards@filenet.com

Budget Rent A Car, Illinois

North America Excellence Awards: Workflow, Finalist

Name: Pat Murray
Title: Project Leader
Phone: 630-955-7076
FAX: 630-955-7799
Mailing Address: 4225 NapervilleRoad
Lisle, IL 60532
email address:

FileNet Corp

Name: Colleen Edwards
Title: Public Relations
Phone: 714-966-3400
FAX: 714-966-3490
Mailing Address: 3565 Harbor Blvd
Costa Mesa, CA 92626
email address: cedwards@filenet.com

Capital Blue Cross

North America Excellence Award: Imaging, Gold

Department/Division: Systems Division
Name: John Reierson
Title: Vice President of Information Services
Phone: 717-541-6060
FAX: 717-541-6072
Mailing Address: 2500 Elmerton Avenue,
Harrisburg, PA 17110

Wang Laboratories, Inc. (Corporate)

Name: Jane Emerson
Title: Director, Strategic Relations
Phone: 508-967-0681
FAX: 508-967-2819
Mailing Address: 600 Technology Park Drive
M/S 01S/470

	Billerica, MA 01821
email address:	jane.emerson@wang.com

Image Consulting Group (ICG—Integrator)

Name:	Maurice Rodriques
Phone:	602-953-8300
FAX:	602-953-8307
Mailing Address:	11811 N. Tatum Blvd., Suite 2700 Phoenix, AZ 85028

Consolidated Edison of New York

North America Excellence Award: Workflow, Silver

Department/Division:	Customer Service, Customer Call Center & Telecommuting
Name:	Ed Glister
Title:	Senior Project Manager
Phone:	212-460-6897
FAX:	212-228-4590
Mailing Address:	4 Irving Place Room 1730 New York, NY 10007
email address:	

Wang Laboratories, Inc. (Corporate)

Name:	Jane Emerson
Title:	Director, Strategic Relations
Phone:	508-967-0681
FAX:	508-967-2819
Mailing Address:	600 Technology Park Drive M/S 01S/470 Billerica, MA 01821
email address:	jane.emerson@wang.com

Wang Software (Field)

Name:	Lisa Ranft
Title:	Account Manager
Phone:	212-644-8526
FAX:	201-309-0956
Mailing Address:	102-21 Shearwater Court East Jersey City, NJ 07305

First Albany Corporation

North America Excellence Awards: Imaging, Finalist

Name:	Edwin Brondo

Title:	Senior Vice President, Chief Administration Officer
Phone:	518 447 8503
FAX:	518 447 8068
Mailing Address:	30 South Pearl Street Albany, NY 12201

Computron Software, Inc.

Name:	Rosemary Mount
Company:	Computron Software Inc
Phone:	770-913 0303
FAX:	770 913 0055
Mailing Address:	Two Concourse Parkway, Suite 600 Atlanta, GA 30328-5585
email address:	103226.206@compuserve.com

Gak Nederland b.v.

European Excellence Award: Workflow, Gold

Department/Division:	Employer Insurance Administration (EIA) Department
Name:	Jaap van Zetten (ASZ)
Phone:	+31 20 6875200
FAX:	+31 20 687 5177
Alternate Contact:	Johan Heetwinkel
Title:	Manager
Mailing Address:	P.O. 8300, 1005 CA Amsterdam, Netherlands
email address:	jzetten@gak.nl

Wang Laboratories, Inc. (Corporate)

Name:	Jane Emerson
Title:	Director, Strategic Relations
Phone:	508-967-0681
FAX:	508-967-2819
Mailing Address:	600 Technology Park Drive M/S 01S/470 Billerica, MA 01821
email address:	jane.emerson@wang.com

Wang Software / Nederland (Field)

Name:	Franklin Willemstein
Title:	Account Executive
Phone:	011-31-345-470707
Mailing Address:	Costerweg 12, 4104 AJ Culemborg, Netherlands Postbus 4, 4100 AA Culemborg, Netherlands
email address:	Franklin.Willemstein@OFFICE.Wang.com

Infocamere, Italy

European Excellence Award: Imaging, Gold

Company Name: Societa' Consortile di Informatica delle Camere di Commercio Italiane per Azioni (Infocamere)
Name: Mr. Pietro Mensi
Title: Vice General Manager and Technical Director
Phone: +39.49.8288111
FAX: +39.49.8288406
Mailing Address: Corso Stati Uniti, 14
35020 Padova (Italy)

Wang Laboratories, Inc. (Corporate)

Name: Jane Emerson
Title: Director, Strategic Relations
Phone: 508-967-0681
FAX: 508-967-2819
Mailing Address: 600 Technology Park Drive
M/S 01S/470
Billerica, MA 01821
email address: jane.emerson@wang.com

Hewlett Packard Italiana

Name: Giovanni Vaccarino
Title: PSO Principal Consultant
Phone: +39.49.
FAX: +39.49.8701856
Mailing Address: Via Lisbona 28, 35020 Padova, Italy

Wang/Italy (Field)

Name: Vincenzo Barba
Title: PSO Consultant
Phone: +39.2.250.701
Mailing Address: Piazza Ghandi 3, 00144 Roma, Italy
email address: vincenzo.barba@OFFICE.Wang.com

Martinair Holland

European Excellence Award: Imaging, Silver

Contact Name: Tom Gasterlaars
Title: Manager Revenue Accounting
Phone: +31 20 601-1942
FAX: +31 20 601-1986
Mailing Address: P O Box 7507
111 ZG Schipol Airport

The Netherlands

FileNet Ltd

Name: Judy Winkler
Phone: 011-44-181-867-6363
FAX: 011-44-181-867-6365
Mailing Address: #1 The Square, Stockley Park
Uxbridge, Middlesex UB11 1FN, England
email address: JudyWinkler/filenet/uk@filenet_uk

Olivetti/Nederland

Name and Title: Evert van Gelder, Account Executive
Name and Title: Richard Fransen, Marketing Manager,
Document and Workflow Management
Phone: 31-71531-9558 or 9515
FAX: 31-71531-5501
Mailing Address: P O Box 11172, 2301 EE Leiden,
The Netherlands
email address: richardf@olivetti.nl

Canada Institute for Scientific and Technical Information (CISTI)

North America Excellence Awards: Imaging, Finalist

Department/Division: Canada Institute for Scientific and Technical Information (CISTI) Document Delivery Division
Name: Bernard Dumouchel, Manager
Mailing Address: Document Delivery, CISTI
Bldg M55
Ottawa, Ont K1A 0S2 Canada
Telephone: 613-993-3969
Fax: 613-952-9112

Logical Software Solutions Corp.

Name: Keith E. Bluford
Title: V.P., Marketing and Sales
Phone: 301-595-2033
Fax: 301-595-2582
Mailing Address: 4041 Powder Mill Road, Suite 300
Calverton, MD 20705
email address: kbluford@lccs.com

NSI Network Support Inc (Integrator)

Name: Clare MacKeigan.
Title: Marketing Manager
Address: 1690 Woodward Drive, Suite 211

	Ottawa, Ontario Canada, K2M1S1
Telephone:	613-226-5571x222
Fax:	613-226-0998

New York City Office of the Comptroller

North American Excellence Award: Workflow, Gold

Department/Division:	Bureau of Law & Adjustment
Name:	Michael Aaronson
Title:	Bureau Chief
Phone:	212-669-4753
Fax:	212-669-2929
Mailing Address:	1 Centre Street, Room 620 New York, NY 10007

Xerox Corp.

Name:	Tom Griger
Title:	Technical Marketing Manager
Phone:	908-245-5388
Fax:	908-245-4094
Mailing Address:	40 Rector Street, New York, NY 10006

Universal Systems, Inc. (Integrator)

Name:	Rod Lustan
Title:	Operations Manager
Phone:	212-859-0698
Fax:	212-480-1295
Mailing Address:	45 Broadway, Atrium, 22nd Floor New York, NY 10006
email address:	Rod_luston@usiva.com

PPP healthcare

European Excellence Award: Workflow, Silver

Department/Division:	PPP healthcare
Name:	Mike Tinsley, Business Project Manager
Address:	Phillips House, Crescent Road Tunbridge Wells, Kent, TN1 2PL England
Phone:	011-44-
Fax:	011-44-1892-503810
email address:	mike.tinsley@pppgroup.co.uk

Mosaix, Inc. (formerly ViewStar)

Name:	John Tarabini
Title:	Manager of Corporate Marketing

Phone: 510-337-2000
FAX: 510-337-2222
Mailing Address: 1101 Marina Village Parkway
Alameda, CA 94501
email address: johnt@viewstar.com

Red Ball Oxygen Company, Inc.

North America Excellence Awards: Imaging, Finalist

Name: Bob Key
Title: MIS Director
Phone: 318-425-3211
FAX: 318-425-6301
Mailing Address: P O Box 7316
Shreveport, LA 71137

Minolta Corp.

Department/Division: Business Info Systems
Name: Ann Marie Marootian
Title: Asst. Director of Marketing
Phone: 201 934-5318
FAX: 201-825-1645
Mailing Address: 101 Williams Dr
Ramsey, NJ 07446
email address: amarootian@minolta.com

Trigon Blue Cross Blue Shield

North America Excellence Award: Imaging, Silver

Department/Division: Medical Claims Processing
Name: Tab Bass
Title: Director Electronic Commerce
Phone: 804-354-4277
FAX: 804-354-7839
Mailing Address: 2015 Staples Mill Road, Maildrop 02A
Richmond, Virginia 23230
email address: tbass@trigon.com

Wang Laboratories, Inc. (Corporate)

Name: Jane Emerson
Title: Director, Strategic Relations
Phone: 508-967-0681
FAX: 508-967-2819
Mailing Address: 600 Technology Park Drive
M/S 01S/470

Billerica, MA, 01821
email address: jane.emerson@wang.com

Image Consulting Group (Integrator)

Name: Maurice Rodriques
Phone: 602-953-8300
FAX: 602-953-8307
Mailing Address: 11811 N. Tatum Blvd Ste. 2700
Phoenix, AZ 85028

Wang Software (Field)

Name: Cliff Gillespie
Title: Account Manager
Phone: 410 828 0550
Mailing Address: 3430 Mark Hall Drive
Marietta, GA 30062

Yarra Valley Water, Australia

Asia Excellence Award: Workflow and Imaging, Gold

Department/Division: Customer Services Group
Name: Gavin Ward
Title: Manager, Sales Channels
Phone: + 613-9874-2122
FAX: +61 3-9872-1353
Mailing Address: Private Bag 1
Mitcham, Victoria. 3132 Australia
email address: gward@yvw.com.au

Wang Laboratories, Inc. (Corporate)

Name: Jane Emerson
Title: Director, Strategic Relations
Phone: 508-967-0681
FAX: 508-967-2819
Mailing Address: 600 Technology Park Drive
M/S 01S/470
Billerica, MA. 01821.
email address: jane.emerson@wang.com

Wang Software (Field)

Name: Therese Whalan
Title: Marketing Manager
Phone: +61 2 9847 7774
FAX: +61 2 9847 7652
Mailing Address: Austlink Corporate Park
2 Minna Close, Belrose, NSW, 2085 Australia

CONTACT INFORMATION

Staffware Pty Ltd

Name: Angela Gregory
Title: Managing Director
Phone: +61 2 9922 4577
FAX: +61 2 9922 4341
Mailing Address: 61 Lavender Street, Level 1
Milsons Point, NSW 2061 Australia
email address: Angela Gregory <agregory@staffware.com>

Contributing Authors

Editor: Layna Fischer

Company: WARIA, Inc. (Workflow and Reengineering International Association)
Title: Chairman
Phone: 954-782-3376
FAX: 954-782-6365
Mailing Address: 3116 N Federal Hwy, Ste 204
Lighthouse Point, FL 33064
URL: http://www.waria.com/waria
email address: layna@waria.com

Priscilla Emery

Company: Association for Information and Image Management International (AIIM)
Title: Sr. Vice President, Information Products and Services
Phone: 301-587-8202
FAX: 301-587-2711
Mailing Address: 1100 Wayne Ave, Suite 1100
Silver Spring, MD 20910
email address: pemery@aiim.org

Tom Koulopolous

Company: Delphi Consulting Group
Title: President
Phone: 617-247-1511
FAX: 617-247-4957
Mailing Address: 100 City Hall Plaza
Boston, MA 02108
email address: tk@delphigroup.com

Connie Moore

Company: Giga Information Group
Title: Vice President
Phone: 540-882-4040
FAX: 540-882-4033

Mailing Address: P.O. Box 3375
Waterford, VA 22910
email address: cmoore@gigaweb.com

Richard A (Ric) Rhodes

Company: DIVA (Document, Imaging VAR Assoc.)
Title: President
Phone: 330-686-7713
Mailing Address: 338 Liberty Rd
Stow, OH 44224
email address: rrhodes@neo.lrun.com

INDEX

The New Tools for New Times series

New Tools for New times: The Workflow Paradigm *(second edition)*

Layna Fischer

This unique and important anthology is the first of its kind to offer a comprehensive guide to the workflow industry and to address the fundamental issues of process reengineering, workflow methodologies and implementation using diverse technologies such as imaging, messaging, forms and databases. The 24-contributing authors, all experts within their respective fields, describe how to make effective changes and offer practical advice on how to implement a BPR program in your company. Detailed case studies.

The book also features a comprehensive vendor directory with over 60 entries, complete with product descriptions and an international industry. Who's Who. Save your time researching dozens of other sources. It's all here. Illustrations, charts, references, appendices, bibliographies, recommended reading, index.

ISBN: 0-9640233-2-6 Softcover $34.95

New Tools for New times: Electronic Commerce

Layna Fischer

Electronic Commerce is the latest anthology in the *New Tools for New Times Series*. The writers— all experts within their respective fields— help you to develop a step-by-step plan for successfully transacting business on-line. They also detail the evolution of the electronic commerce industry through electronic data interchange (EDI) to the Internet, workflow and the World Wide Web, offer practical solutions and a peek into the future of electronic commerce.

Quality softcover, 378 pages. llustrations, charts, references, appendices, bibliography, recommended reading, index. 6"x9".

ISBN 0-9640233-3-4 $29.95

Order Form

The Workflow Paradigm, Second Edition (Fischer) $34.95
Electronic Commerce (Fischer) $29.95
Excellence in Practice-Innovation and Excellence in Workflow Imaging (Fischer) $50.00

SHIPPING INFORMATION:

Name:______________________________
Title/Occupation:________________________
Company:____________________________
Address:____________________________

Phone: ______________ Fax: ______________
Email:______________________________

PAYMENT INFORMATION:

_____COPIES @ $_____ = __________________
(make copies of this page for additional titles.)
_____FL State tax 6% __________________
_____shipping (see below):__________________
subtotal:__________________
total: __________________

Check/money order (in US Dollars drawn on a US Bank to Future Strategies Inc.)

Visa MasterCard Amex Diner's Club *(circle one)*

Card Number: ____________________________
Expires on: ____________________________
Signature: ____________________________
Date: ____________________________

Mail or fax this order to:

Future Strategies Inc., Book Division
3116 North Federal Highway, Lighthouse Point, FL 33064 USA
Tel: +1 954 782 3376 Fax: +1 954 782 6365 email:waria @gate.net

*AIRMAIL SHIPPING CHARGES **PER BOOK**: USA Priority Mail $3.50; Canada/Mexico $6.00; UK/Europe $11.00; Pacific Rim $13.00; Africa/South America $17.00 Ask for quote on bulk orders.
(Distributors/Bookstores/Libraries/Educational Institutions - please call for special discounts and shipping price schedule.)